Adva

Advanced Language Learning:

The Contribution of Halliday and Vygotsky

Edited by Heidi Byrnes

continuum

Continuum
The Tower Building
11 York Road
London SE1 7NX

80 Maiden Lane
Suite 704
New York, NY 10038

British Library Cataloguing-in-Publication Data
A catalogue record for this book is available from the British Library.

ISBN: 0–8264–9071–9 (HB)
ISBN: 978–0–8264–4308–3 (PB)

Typeset by RefineCatch Limited, Bungay, Suffolk
Printed and bound in Great Britain by
the MPG Books Group

Contents

List of Contributors

Heidi Byrnes, Georgetown University

Alice Caffarel, The University of Sydney

M. Cecilia Colombi, The University of California, Davis

Cori Crane, University of Illinois at Urbana Champaign

Parastou Feiz, The Pennsylvania State University

Dessislava Ivanova, The Pennsylvania State University

James P. Lantolf, The Pennsylvania State University

Christian M. I. M. Matthiessen, Macquarie University, Sydney

Nick Moore, Etisalat University College, United Arab Emirates

Marianna V. Ryshina-Pankova, George Mason University

Mary J. Schleppegrell, The University of Michigan

Susan Strauss, The Pennsylvania State University

Merrill Swain, The University of Toronto

Kazuhiro Teruya, The University of New South Wales, Sydney

James V. Wertsch, Washington University, St Louis

Xuehua Xiang, The Pennsylvania State University

List of Figures and Tables

Figures

Tables

What kind of resource is language and why does it matter for advanced language learning?

An introduction

Heidi Byrnes

This collection brings together three areas of inquiry within language studies that have thus far not been considered together: first, a particular theory of language, systemic functional linguistics (SFL), as laid out over roughly the last four decades by M. A. K. Halliday and his followers, primarily in Australia; second, a particular theory of the nature of human cognition and learning in relation to language, sociocultural theory (SCT), as originally developed in little more than a decade in the mid-twenties and early thirties of the last century by Lev Vygotsky in the former Soviet Union; and, third, a particular area of language use and development, namely second or foreign language (L2) development by adult learners at 'advanced' levels of ability, which has recently come to the fore in professional discussion. As the title of the volume indicates, within that triangle advanced language learning is in focus or, in reverse, advanced L2 capacities provide the lens through which links to SFL and SCT will be explored, with the intention of illuminating the nature of those capacities and facilitating their development within an educational context.

My reflections in this introduction will follow these steps: I begin with an exploration of central assumptions, insights and constructs in both SFL and SCT in order to probe them for their potential to illuminate aspects of L2 advancedness and teaching and learning toward advanced capacities. As it stands, neither theoretical framework has explicitly addressed that level of language learning, though much in them invites its exploration. Coming from the other side, to date advanced instructed L2 learning itself has received only scant notice in the language profession.[1] Portraying a professional context that has largely ignored advanced L2 learning and also a sociopolitical and educational context that increasingly demands it urgently are therefore necessary steps in order to further locate possible future discussion. As a way to stimulate it, the next section offers an exploratory look at the potential for reconceptualizing advancedness in light of SFL and SCT theory in the area of grammar, lexicon and text. I conclude by raising some issues for advanced L2 learning in light of these two theoretical frameworks.

Let me clarify a bit further the volume's focus. It is advanced *foreign* language learning in an instructed setting as contrasted with *second* language learning in a naturalistic environment. That choice comes about not because the designation is clear-cut or easily maintained in opposition to naturalistic second language learning. Indeed, contributions in this volume point to an increasingly prominent 'third space' in language instruction, namely a combination between various forms of immersion/residency experiences and formal instruction, but with the distinguishing characteristic of *actually thematizing* the special role of instruction, rather than simply being aware of its existence – or ignoring its larger context. Thus, purely instructed foreign language learners (Caffarel, Moore, Ryshina-Pankova, Teruya), heritage speakers (Colombi, Schleppegrell), students in various bilingual environments (Swain) or sojourning ESL students in the United States who may or may not return to their home countries (Strauss *et al.*) are all of concern in this collection as they seek to enhance their language capacities toward academic levels of performance, one way to describe 'advanced' abilities.

Rather, the focus was chosen because it can bring into strong relief two issues in language education that will have to become an imagined reality if advancedness is to be a viable educational goal. The first is the assumption that advanced L2 abilities can, even must, become a reasonable learning outcome for entire language programs and not only for individual 'gifted' learners, even when learners are linguistic adults, most particularly in college-level contexts. This assumption would apply not only in countries like the United States where tertiary education already carries a considerable load in creating a population with meaningful second/third language abilities. Increasingly, it will also apply elsewhere because the so-called less commonly taught languages, usually non-Indo-European languages that are not part of the secondary school curriculum, will have to be acquired beyond that educational level. Naturally, the demands for efficiency and effectiveness will only rise. Technology is often invoked in order to come to the rescue. But the question that needs answering is not whether but how to use it appropriately. Moore (this volume) offers a way of linking the capacities of modern computers to analyse vast databases for lexicogrammatical features that have a high probability of occurring in certain professional and academic registers and using that insight for curriculum construction that would address just that concern for efficiency. Phrased more broadly, the question is this: is efficiency and effectiveness achievable if, perhaps only if, a broad conceptualization of language as a meaning-making system is vigorously asserted?

The second and related assumption is that, contrary to popular judgements of classrooms as severely impoverished environments for any language learning, the instructed setting may in fact be a particularly beneficial, even necessary environment for the attainment of advanced capacities. This stands on its head long-standing beliefs about immersion or study abroad settings being nearly the only way to reach advancedness. Of course, no one would dispute the benefits of an L2 cultural sojourn. But with regard to attaining advanced levels of L2 ability that go beyond the perennially nebulous notion of 'fluency' (for a good discussion, see Freed *et al.* 2003) the operative factor may not so much be location, i.e., study abroad versus continued study in the home country, as the nature and breadth of learning opportunities. Such opportunities would include a carefully considered range of texts that would make certain situated textual demands on the learners and would enable them to develop a differentiated meta-awareness about the consequences of choices made in and for construing particular communicative situations. In that case, a well-designed instructional program might be able to offer a richer palette of acquisition-attuned textual varieties and tasks, along with scaffolded learning environments, than would many a study-abroad sojourn.

One closing observation for these prefatory comments: I will not provide the standard separate summary of chapters of the volume. Rather I will weave their diverse foci into the narrative itself. I have confidence in the readers' ability to interpret the nature and significance of the individual papers for a better understanding of advancedness and how we might foster it in our educational practice; and I have equal confidence in the authors' ability to have made their case in an enticing and persuasive fashion. Perhaps the kind of co-location of heretofore largely separate intellectual pursuits in the three areas that the volume links can then evolve in the readers' mind into collocations that enliven not only their own thinking about advanced learning but can become within the profession what Wertsch (this volume) calls a 'dynamic form of dialogic energy'.

Motivating a nexus between SFL, SCT and advanced L2 learning

Even a cursory glance through the professional literature confirms the link between SFL, SCT and advanced L2 learning to be an unusual configuration. To some, such novelty might make it suspect right from the start; others might welcome it as an instance of novel transdisciplinary thinking. However that is viewed, the proposal invites at least general consideration of each component for already existing and potential links to one or the other 'end points'. At the same time, some background is necessary for why the total configuration might recommend itself.

SFL and its affinities

For SFL the affinities to SCT and advanced learning are striking. In the first case, they have been explicitly stated; in the second case they arise more through interpretive projection.

Assuming that instructed advanced L2 learning is at heart an educational issue, the following observation made by Christie and Unsworth (2005) is significant: among the major theoretical frameworks in linguistics Halliday's SFL is surely the most explicitly education-oriented. Furthermore, it is so not as an afterthought in which an existing fixed theory is 'applied' in or to an educational context. Rather, according to Halliday's own judgement, the expansive body of theoretical, empirical and educationally oriented scholarship and practice of SFL often arose in response to external, often educational, challenges and opportunities (1996: 21; 2006). This is true both in Halliday's publications and in the published and educational work of many of his key followers (e.g., Christie, Hasan, Lemke, Martin, Matthiessen, Mohan, Rothery, Unsworth, Veel, Williams). Thus, various foundational assumptions in Halliday's introduction to functional grammar – in its various revisions still the most compact and complete treatment of systemic-functional theory; Halliday 1985/1994, now Halliday and Matthiessen 2004 – can be traced back to assumptions that had been developed earlier on within the British educational context of the nineteen sixties and early seventies. For example, Halliday argues that a language is 'functional in the sense that it is designed to account for how the language is **used**' and that 'the fundamental components of **meaning** in language are functional components' (1994: xiii, original emphases). That strong statement about language in use as being all about meaning-making is elsewhere elaborated in terms of an explicitly language-based approach to any meaning-making and knowing, including, quite expressly, coming to mean and know in the context of learning as that is practised in educational contexts (Halliday 1993).

In that article, Halliday offers an expansive ontological view that spans from the very beginning of a child's language-non-specific protolanguage to the complex literacies of the adult, which are characterized by a multi-layered capacity to construe experience in language. On the one hand, it explores at considerable theoretical depth the relation between the semiotic tool language and the human capacity to mean; on the other hand it draws on much earlier insights gained by educational practitioners, namely that 'educational failure is primarily *linguistic* failure' (series editor introduction to Halliday 1973: 3, original emphasis).

Halliday states the observed trajectory like this: human beings are quintessentially creatures who mean and therefore any theory of language must account for how it is that language enables us to mean and to know. Beyond that initial way of meaning and knowing through language in the pre-school years, there is a decided shift, from semiosis in speaking to semiosis largely based on writing that characterizes education. This, in Halliday's terminology, is a shift from commonsense ways of knowing to new forms of knowledge that are distinct and distinctive for educational knowledge. He describes it in terms of the following six steps:

- from interpersonal orientation (language as action) to include experiential orientation (language as understanding)
- from dialogic mode to include monologic mode
- from the deictic centre ('you-and-me, here, now') outwards to include 'other persons and objects', 'other times', 'other places'

- from entities that are concrete and perceptual to include entities that are institutional or abstract
- from simple categories ('common terms') to include taxonomies of categories
- from generalization to include prediction, reasoning and explanation (Halliday, 1999a: 80)

And he adds: 'These are preconditions for learning to read and write and for acquiring systematic knowledge under instruction' (*ibid.*).

Lest this appear to be no more than the familiar 'expanding circles' of experience notion of much of L1 education in the primary grades or the progression in functional-notional as well as communicative approaches in L2 teaching from self to community to larger world, it is critical to emphasize that a language-based theory of knowing and learning investigates the nature of the language resources needed for enabling such ways of knowing, rather than focusing nearly exclusively on the settings (e.g., a visit to the zoo) or the content of the imagined or real communicative events (e.g., reporting on a science project or summarizing the plot of a story). And here, SFL has, perhaps, made its most important contributions, through explicating, particularly by way of the construct grammatical metaphor, how grammar reconstrues experience from commonsense ways of knowing to metaphorical ways of knowing and understanding. I will address these notions further in subsequent sections of this paper but refer readers particularly to Matthiessen's careful treatment and also to a number of contributions that use this framework (Caffarel, Colombi, Crane, Ryshina-Pankova, Schleppegrell and Teruya).

As a result, the perception of language in education differs dramatically from current SLA research and educational practice: 'Language is not a **domain** of human knowledge . . . Language is the essential condition of knowing, the process by which experience **becomes** knowledge' (1993: 94, original emphasis). Many deeply ingrained habits of mind, both in L1 education, but considerably more insidiously in L2 education, are thoroughly undermined by that statement. Among them are such long-standing practices as separating language from content, or form from meaning, or separating syntax from discourse from semantics from pragmatics. All miss the point made by Halliday, even as they assert their awareness of the intimate relation between language, thought and culture and are eager to *add* those components to their language instructional proposals. But language learning is not a skill that can be enhanced through decontextualized and content-less learning strategies. More specifically, the No Child Left Behind legislation notwithstanding, reading is not a skill to which content can be added once it is sufficiently well developed. If that were understood then the current vigorous attempts to improve students' reading abilities would not crowd out curricular content, such as history or social studies or literature, as many teachers and school districts in the United States report. To the extent that educational practice does not attend to the kind of expansion of linguistic resources necessary for expanding understanding in the various subject matter areas of education through enabling learners to participate in this reconstrual of reality through language it severely limits their ability to mean – in L2 just as in L1.

With an understanding of language learning as learning how to mean in new and different ways SFL scholars became engaged with diverse educational issues in diverse school settings. They developed a genre-based pedagogy that carefully observed the subject-specific demands in different disciplinary areas being made of students (see e.g., contributions in Christie and Martin 1997; much of Martin's work, as cited throughout the volume). Taking the perspective of an educational linguist, Rothery (1996) provides an instructive summary of that work: from the analysis of genres that had the greatest frequency in primary schools – from picture description to story genres (e.g., observation, recount, narrative) to the factual genres of report, procedures and exposition/explanation; to the preferred genre of secondary schooling, the factual genres that involved procedure and persuasion, the story genres of news story, exemplum, moral tale or fable, narrative, recount and observation; and the genres that invited a response in the form of a review, an interpretation or a critical analysis (see also Matthiessen's topography of genres, this volume).

Despite surface similarities this kind of 'needs analysis' differs markedly from recommendations in standard SLA textbooks (e.g., Brown 1995, particularly Chapter 2, Needs Analysis). Thus, Brown leaves open to institutional decision-making the kinds of 'points of view' it wants to take with regard to how or whether to link in any principled way linguistic content, situation and language. He states that, of the possible philosophies and dichotomies for dealing with these parameters, 'none is advocated over the others because the decisions about which roads to follow in a particular language program depend on the personalities and institutions affecting that program' (1995: 42). It is difficult to see how that conclusion, even when accompanied by the admonition that early decisions about these matters are critical because 'they can save an enormous amount of backtracking, wasted energy, and frustration' (*ibid.*), can amount to a well-theorized notion of curriculum construction using needs analysis. By contrast, SFL researchers were able to develop through a detailed genre-analysis a critical genre-based pedagogy that strives to develop a critical literacy in students. (For a recent translation of these insights into the American context, see Schleppegrell 2004; Crane, this volume).

The aim is a competent level of literacy on the part of learners that, crucially, involves awareness of the meaning-making consequences of different linguistic resources at all levels of language, from the lexicogrammar to the staging of arguments in texts, both oral and written. Once again, its animating foundation is a language-based theory of learning that occasions theoretically insightful discussions of literacy (e.g., Hasan 1996b) alongside innovative curricular and detailed pedagogical initiatives in literacy education (e.g., Martin 1997, 1999, 2000).

I conclude this section with something of a reflective coda: why has the expansively developed theoretical framework of SFL, along with its expansive educational work, found so little resonance in the U.S. American context of language theorizing and also in pedagogical reflection, a state of affairs that is only recently beginning to change (see particularly the efforts by Colombi, Mohan and Schleppegrell and their co-workers)? However one wishes to 'explain' that phenomenon, it surely reflects the dominance in U.S. American academic inquiry of decontextualized and ahistorical theorizing, even in such a highly context-dependent and socially construed semiotic environment as language, in other words, a preference for form and structure over function and meaning. American structuralism, on which most language pedagogy and SLA research continues to rely, and universal grammar approaches championed by Chomsky are the most well-known exponents of such thinking. The depth of that privileging, if not intellectual isolation, is eloquently demonstrated when proponents of a dramatic shift in the direction of a meaning orientation in language, such as Fauconnier (1997) and Langacker (1998), make no reference to systemic functional grammar even though they assert that language must be studied in its discursive context and is intimately implicated in reasoning and social communication.[2] At the same time these issues have preoccupied Halliday's SFL for well over four decades.

Similarly, from the SLA research and practitioner side, the fanfare sounded by the seminal article by Firth and Wagner (1997) that a positivistically and psycholinguistically driven SLA enterprise was increasingly unable to explicate central processes in SLA has begun to enlarge ontological and empirical preferences in SLA (for a recent summative statement, see Block 2003). Though markedly different in theoretical apparatus and focus, Lantolf's work is part of that trend, which foregrounds a social and contextual understanding of language, language acquisition research and pedagogical practice. Nevertheless, the language profession continues to accept as given not only the conceptual umbrella of 'scientific facts' as corroborated in SLA research within that framework – the field's dominant 'paradigm' – even though it is unable to contribute substantively to notions of advancedness; it rests secure as well under the accompanying practice-oriented umbrella that comes in the form of a flood of materials and teacher education efforts, particularly in the area of ESL/EFL teaching.[3]

Whatever one might ultimately take as reasons, the result is a complete sidelining of issues of advancedness beyond 'more and better'.

The sociocultural turn: Vygotsky's theory of mind and language

Especially through the efforts of Wertsch in psychology (e.g., 1985, 1998, 1990, 1991, 2000 and this volume) and Lantolf in second language acquisition research (e.g., Lantolf and Appel 1994; Lantolf 2000a and b, 2006 and this volume) and their collaborators and followers, SCT with a strong Vygotskian orientation has gained recognition within applied linguistics research and in the educational practice of teacher training and pedagogy (e.g., Hall

2002). Two summaries of SCT from an SLA perspective (Lantolf 2000a, 2006) highlight the following features:

Perhaps the core construct for SCT is that of mediation and that applies particularly to language learning. Lantolf states: 'Sociocultural theory holds that specifically human forms of mental activity arise in the interactions we enter into with other members of our culture and with the specific experiences we have with artifacts produced by our ancestors and by our contemporaries' (2000a: 79). Building on that centrality for mediation, Vygotsky further argues that higher forms of human mental activity are dependent on symbolic tools that cultures have developed over time, with language being the most critical of these. Coming from a psychological perspective, his interest was not so much the fully formed system of the adult, but how those mental capacities are gradually formed in the human person. In other words, the *process* was likely to be more revealing of the organization of mental activity than merely observing the *product*, a stance that is referred to as the *genetic method* because of its emphasis on the history of concept formation.[4] It is here that the concept of the *zone of proximal development* becomes critical, not as a place or a context, but as a

> dialectic *unity* of learning-and-development, or more appropriately, learning-leading-development . . . In this unity, all uniquely human forms of higher mental activity, including thinking, planning, voluntary memory, voluntary attention, creativity and control of semiotic systems (especially language), arise in the interaction between children and other members of a culture during ontogenesis. (Dunn and Lantolf 1998: 420, original emphasis)

Taking education to be a privileged environment for such interaction, this context itself becomes a way of 'creating' development, rather than just a way of responding to it. In Dunn and Lantolf's phrasing, 'instruction and learning do not ride on the tail of development but instead blaze the trail for development to follow' (1998: 419). Expanding that notion beyond the customary beginning and intermediate levels of instruction would challenge the profession to imagine its repercussions within the increasingly language-mediated activity of education and its demand for advanced L2 abilities.

A second major concept is that of internalization, which addresses the movement from social ways of knowing to increasingly internal ways of knowing, a development where imitation and private speech play a critical role. While earlier studies investigating L2 learning have examined the use of private speech as a way of regulating task completion, more recent work has looked at the role of private speech for internalization of culturally and linguistically shaped concepts as well as the interface between speech and gesture as a particular form of mediation that provides insight into the use of conceptual metaphors (Lantolf 2006, also this volume). Lantolf relates his interest in private speech and conceptualization to Slobin's notion of thinking for speaking (1996) and, more generally, cognitive linguistics with a strong semantic orientation (e.g., Fauconnier, Slobin, Talmy, Turner).

Significant for our discussion is that these two theoretical approaches have been extended into issues of L2 advancedness. Indeed, a number of the papers in Byrnes *et al.* (2006), particularly those by von Stutterheim and Carroll (2006), Carroll and Lambert (2006), and Behrens (2006), pursue exactly that route and reach provocative conclusions regarding the nature of advancedness. But they do so within a textual environment that both complicates the nature of what it means to think for speaking in a second language and also explicates many phenomena of advanced language learning. In the process, that research not only strongly reasserts Halliday's insistence (1999b) that texts are how language relates to social processes in a principled way; presumably, therefore, it is also in texts that we need to seek foundational characteristics of advanced capacities. It also reorients some of the long-standing claims associated with ultimate attainment by adult L2 learners.

Probing the complementary contribution of SFL and SCT to advanced L2 learning

Based on this admittedly sketchy outline of the key tenets of SFL and SCT I now explore how they themselves see their relationship. Given my earlier statement about the non-familiarity of North American theorizing and practice in the language field, that connection is established primarily by SFL and, not coincidentally, by a scholar in educational research. Three vignettes, chosen for their link to advancedness, will need to stand in for diverse debates: a theoretical concern, a contextual re-reading of Vygotsky's alleged focus on the word level, and the claim for complementarity between the two theories presented by Wells (1994).

Scholars who work within the SFL framework show considerable intellectual engagement with Vygotsky's theory of mind (see particularly Hasan 1996c, 2005a and b; Martin 1999; Williams 2005), viewing it as highly compatible with SFL along several lines: in terms of its genetic, that is, its developmental orientation to understanding human cognition as an ontogenetic and a phylogenetic phenomenon that has both biological and social foundations, thereby overcoming the opposition between these two realms; in terms of its understanding of mental functions, particularly higher mental functions, as being mediated by 'artificial stimuli,' particularly language; and in terms of understanding higher mental functions as 'always sociogenetic. Their nature is social not in the simple sense that they are tool-mediated and tools are social *in themselves*. Rather the cultural, interactional process is a necessary element for conceptualizing something as a mediating means, no matter whether these means are concrete or abstract' (Hasan 2005a: 110). The contribution by Strauss *et al.* (this volume) exemplifies these points particularly well.

At the same time, Hasan also notes serious lacunae in Vygotsky's framework, among them an inability to address how particularly valued forms of semiotic mediation come about, how differences in semiotic mediation and human consciousness come to exist and how the theory handles valuation of variants itself. Because advanced learning is surely about an expanded capacity for enacting situated variability of language use in particular contexts of situation (see Malinowski 1935), that limitation must be addressed.

Wertsch speaks to just that issue when he invokes Bakhtin's concepts of genre and dialogism in order to explicate the link between psychological and social institutional phenomena (1985; see also Hasan's careful analysis of Vygotsky and Bakhtin in 1996c, 2005a and b). In his contribution to his volume Wertsch further explores the essential dialogism not only of utterances in 'local dialogue', but of any utterance in relation to 'the generalized collective dialogue': knowing a language at the advanced level could then be described in terms of being able to link it in both directions, from the instance to the system and from the system to the instance. Similarly, Bakhtinian along with Vygotskian perspectives are invoked by Strauss, Feiz, Xiang and Ivanova (this volume): in their research on advanced ESL writing they find that the dialogic and collaborative co-construction of meaning over model texts can become a key mediational tool for learners in a classroom setting to internalize ways of thinking textually.

Although Hasan recognizes Bakhtin's enormous contribution to a link between the individual and the social in language through the notion of speech genre, in the end she concludes that it is insufficiently developed inasmuch as it has no developed theory of social context, such that

> one, it would explain the principle whereby the immediate social situation is related to social milieu; and two, it would specify the composition of social situation itself making salient those of its significant elements which are relevant to the understanding of the linguistic facts as they impinge on utterances and utterance types. Ideally, the theory would attempt to specify the *principles* by virtue of which the elements of the social situation happen to be related to the wording and the compositional structure of the utterance (types). (Hasan 1996c: 169)

In the second category, that of alleged specific shortcomings of Vygotsky's framework, I refer to one that has repeatedly been noted (e.g., Hasan 1996c; Lucy and Wertsch 1987) but that might also resolve itself with careful exegesis in context. Thus, Williams concludes that 'word' for Vygotsky did not refer to individual lexical items; instead, from the context of its occurrence as a way of understanding how children's attention becomes directed, it is more akin to 'thought being primarily formed by language in use' (2005: 288). Similarly rejecting a restricted interpretation of Vygotsky's term 'word', Lantolf (this volume) puts it into proximity with his notion of 'languaculture', a way of reuniting language and culture in a necessary, rather than a merely casual link.

Finally, one of the earliest explorations of Halliday's proposal for a language-based theory of learning in relation to Vygotsky's theoretical framework comes from Wells, an educational researcher who is particularly strongly engaged in teacher education (1994). In response to Halliday's earlier-referenced 1993 article, he offers a specifically education-oriented analysis of both. Comparing Halliday and Vygotsky's key assumptions and representations (e.g., a genetic approach to language, language and social activity, learning language as appropriating culture, language and intellectual development, language and thinking in school, the educational consequences of sociosemantic variation), he concludes:

> A comprehensive language-based theory of learning should not only explain how language is learned and how cultural knowledge is learned through language. It should also show how this knowledge arises out of collaborative practical and intellectual activities and, in turn, mediates the actions and operations by means of which these activities are carried out, in the light of the conditions and exigencies that obtain in particular situations. Furthermore, such a theory should explain how change, both individual development and social and cultural change, occurs through the individual's linguistically mediated internalization and subsequent externalization of the goals and processes of action and interaction in the course of these activities. (Wells, 1994: 84)

He is concerned that Halliday's near exclusive focus on language as a meaning-making tool comes at the expense of other forms of learning, such as meaningful action. In the context of advanced L2 learning it is not yet clear which approach is more conducive to being realized in institutional contexts: Vygotsky's approach with its higher potential for being linked to activity that is not necessarily language-based (and, by implication, to contemporary activity theory) or Halliday's elaborated understanding of learners' expansion of their meaning potential, primarily by means of language realizing discipline-based knowledge.

Advancedness in current professional discussion

I commented at the outset of this paper on the near-absence in professional talk of the notion of instructed language learning to advanced levels of ability and thus far have probed that fact nearly exclusively from the theoretical side. But one can surely argue as well that such an interest should not need to be justified with theories of language: programmatic foci typically do not arise from theoretical constructs. In other words, one needs to look a bit farther afield to understand the remarkable restriction of vision to essentially the beginning and intermediate levels of instruction.

I have attempted to do this over a number of years from several perspectives. Taking a larger societal viewpoint (Byrnes 2004) I have argued that what might, in the past, have been a privileged enterprise necessary and suitable for only a few, namely the acquisition of language capacities that can be used in academic, institutional and professional contexts, as contrasted with primarily personal and social contexts, is now 'beyond option or privilege'. Moreover, in a globalized environment in which the sovereign nation state, including its construction through national languages, is being reshaped, many of the assumptions that have undergirded mostly beginning- and intermediate-level language instruction are in any case being questioned severely, thereby at the very least attenuating their validity (e.g., single and/or fixed norms, canonical texts, separation of nativeness and foreignness, structurally rather than functionally oriented notions of language and language learning).

From inside the profession I have explored the phenomenon from the curricular standpoint (1998) highlighting historical, structural and general intellectual concerns that combined with an environment in SLA research to make genuine curricular thinking over longer instructional periods virtually impossible. That, of course, also made thinking of and about advancedness virtually impossible, for the simple reason that language learning to competent levels of performance is a process of some duration. A second factor contributing to the silence surrounding L2 advancedness may lie in the near-complete transformation of foreign language departments as literature departments into cultural studies departments. For all its claims of the centrality of culture in language, that change, very much under the influence of postmodernist theorizing (often translated from French texts!), actually reduced inquiry and teaching in FL departments to a largely language-less sociological phenomenon. When, in addition, English often became the language of instruction, that development precluded any further discussion in FL departments of thought as being 'language-based' in a meaningful fashion, and, by extension, any meaningful reflection on advanced L2 learning (Byrnes 2002a).

The incongruity could hardly be greater: educational contexts that have the most solid institutional anchors – by no means a given for 'language teaching' – and that have the greatest intellectual, programmatic and pedagogical interest in finding and then vigorously affirming the link between language, thought and culture are nearly helpless to engage in such thinking (see Byrnes and Kord 2002 for a dialogue on the matter between an SLA and literary researcher). Nor does it end there. Targeting foreign language supervisors who direct introductory and intermediate instruction, papers in Byrnes and Maxim (2004) attempt to overcome the intellectual and structural limitations of that position. By using L1 literacy scholarship as well as early explorations of SFL approaches to genre, they strive to place language acquisitional concerns alongside the standard content concerns at the upper levels of programs. It is no accident that only one voice, that of Swaffar (2004), represents the concerns of a language-oriented literature scholar.

While the ACTFL/ILR proficiency framework continues to inform discussion about advancedness (e.g., the majority of contributions in Leaver and Shekhtman 2002; but see Byrnes 2002b), Hallidayan, cognitivist and Vygotskian perspectives are increasingly shaping a different notion of 'advancedness' in the U.S. American context. Thus, Schleppegrell and Colombi's edited volume (2002) presents central conceptual tools available within SFL and applies them to the development of academic literacy in diverse educational contexts, including ESL, heritage learners and issues arising with bilingualism and the preparation of teachers. How those insights, particularly the notion of genre-based literacy, might reorient programs reaching toward advanced L2 development, including programs that favour 'content-based instruction', is the focus of a number of my own publications (Byrnes 2004, 2005a and b, 2007; Byrnes and Sprang 2004; Byrnes *et al.* 2006a). Finally, a cognitive linguistic perspective on advancedness is prominent in Byrnes *et al.*, (2006b) and the implications of targeting advancedness for research methodologies are explored in Ortega and Byrnes (2007), where long-term trajectories are seen as necessary both for capturing the nature of L2 capacities and their gradual emergence over time.

I have recounted these developments in order to describe the professional landscape within which discussion about L2 advanced learning would be taking place. The next section is a necessarily brief but, I hope, nonetheless suggestive consideration of L2 advancedness in light of SFL and SCT theory.

What kind of resource is language? Translations into advancedness

This entire collection of papers and, therefore, my introductory remarks have built on my strong sense that the notion of L2 advancedness remains to be specified in a way that captures key aspects of the phenomenon; I have presented SFL and SCT theory as particularly suited to the task. I have made this argument without ever having specified what 'advancedness' refers to in the first place. The reasoning goes something like this: we know it when we see it, though we may have difficulty defining it, and may have disagreements about the categories that would need to be included and the degree of importance we would ascribe to them.

Naturally, this concluding section can do no more than sketch out, in the broadest of brushstrokes, some potential ramifications of situating advancedness within the nexus of SFL and SCT. The papers in this volume, the majority of which overtly take an SFL perspective while making many an assumption that accords with Vygotskian themes, provide what enticing details are possible within the confines of single papers within a single volume. In order of occurrence they have the following foci: Matthiessen presents an expanded discussion of SFL as a descriptive framework whose very shape as a systemic-functional grammar has what Halliday calls a 'metaphoric' relationship to the object of its description (1996: 19); Teruya explores how the capacity to develop logical relationships within the experiential world depends on a sophisticated awareness of the diverse linguistic means that realize them in Japanese; turning to writing, Schleppegrell specifies the linguistic qualities of English-language expository texts, Colombi uses the central construct of grammatical metaphor as a way to enable Latino students to develop academic levels of ability, and Ryshina-Pankova focuses on the textual function in order to specify the nature of coherence and cohesion in advanced L2 German writing; reminiscent of Swain's classroom with younger learners of French, Caffarel uses talking about language as a resource for meaning-making in language and learning French at the college level; and Crane proposes a model for a foreign-language curriculum that is based on the construct of genre as a particular configuration of registerially and sequentially marked textual qualities.

By comparison, my own reflections are at once more speculative and address broader concerns.[5] Even if one acknowledges that the definition of 'advancedness' is likely to be very much in the eye of the beholder, from the practitioner perspective – and, theoretical interests notwithstanding, this is a critical perspective – three areas are nevertheless likely to stand out as meriting particular attention: 'grammar', 'vocabulary' and 'text'. These are, of course, pretty much the staples of our profession. That means they have been around for being shaped discursively for a long time. As different perspectives on them now become available, one should be quite clear that even if some of these phenomena, particularly 'text', have received only peripheral attention, they have nonetheless been situated within existing frameworks. These have provided the cognitive metaphors with which the profession has traditionally thought about all its work – no matter where it takes place and who engages in it, though with a noticeable slant toward beginning and intermediate levels of acquisition. Furthermore, over the last two decades or so, these cognitive metaphors have both defined intellectual space and naturalized it in our 'thought collective' (see endnote 3). Concepts like input, output, interaction and negotiation for meaning have meant certain things; likewise, grammar, form and meaning have meant certain things; and the classroom with its feedback and recasts, that are uptakes or not, has also meant certain things. Because these are not innocent references to pre-existing facts but come with powerful territorial claims in the intellectual realm and therefore with regard to professional identities, they will be difficult to dislodge or to discard.

If, then, we follow through on the assumption that our ways of knowing are linguistically driven, then a particular challenge for developing a new orientation will be a languaging challenge: as Swain demonstrates so well with her students, talking it through with the resources the *new conceptual language* makes available is going to be essential to new ways of knowing. Perhaps – and this is a wish that I hope is not merely wishful thinking – since those new ways of knowing pertain to a *new area of language learning*, namely advancedness, the project can in fact succeed.

Meaning with grammar: a semiotic perspective

Of the many perspectives on advancedness SFL makes available, perhaps none is more insightful on a deeper level and applicable in classrooms than Halliday's understanding of grammar as a 'meaning-ful' resource. I have chosen three perspectives on grammar: in terms of its semiotic foundation, as the consequence of a privileging of the paradigmatic over the syntagmatic axis of language description, and in terms of the notion of construal.

Assuming the semiotic character of language Halliday (1996) differentiates higher order semiotic systems against primary semiotic systems. On the one hand this allows him to echo the trajectory we have already traced with

Vygotsky, namely from the physical to the biological to the social to the semiotic, that is at the heart of any meaning-making. On the other hand, it enables him to identify grammar as the key distinguisher between these two kinds of systems. For while any semiotic system, e.g., gestures, is inherently based on social realities and has the capacity to 'mean' and is 'functional' in that sense, only higher order systems are endowed with a grammar. From an evolutionary standpoint that capacity results from what Halliday refers to as a 'deconstructing of the original sign and reconstructing it with the content plane split into two distinct strata, semantics and lexicogrammar' (1993: 6). From a meaning standpoint this grammar/lexicogrammar

> is an entirely abstract semiotic construct that emerges between the content and the expression levels of the original sign-based primary semiotic system. By 'entirely abstract' I mean one that does not interface directly with either of the phenomenal realms that comprise the material environment of language. The expression system (prototypically, the phonology) interfaces with the human body; the (semantic component of the) content interfaces with the entire realm of human experiences; whereas the grammar evolves as interface between these two interfaces – shoving them apart, so to speak, in such a way that there arises an indefinite amount of 'play' between the two. (*ibid.*, 6)

In that case, grammar becomes a privileged part of language: as a critical component of our ability to mean it is in constant dynamic interface between the ever-changing material world and our semiotic world and gives language its meaning-making energy (cf. Halliday, 1996: 4).

From the standpoint of language analysis, that meaning focus requires a dramatic shift away from the syntagmatic axis of syntactic rules, the dominant preoccupation of structuralist and universal grammar theorizing, to the paradigmatic axis. Halliday characterizes his decision to favour the paradigmatic axis unequivocally as having been motivated by the fact that grammar or, more precisely, grammatics as the field of inquiry that describes the phenomenon of grammars in languages, must be all about exploring the central quality of language, namely its function for enabling us to make meaning in the world. What is remarkable about that choice is that he traces it back to the sixties, where the particular challenges facing grammatics were

> '**computational** (machine translation), **educational** (first and second language teaching; language across the curriculum); **sociological** (language and cultural transmission in Bernstein's theoretical framework . . .; **functional-variational** (development of register theory) and **textual** (stylistics and analysis of spoken discourse. All these tasks had in common a strong orientation toward **meaning**, and demanded an approach which stretched the grammar in the direction of semantics. (Halliday 1996: 20–21; emphases added)

It hardly needs a reminder that American linguistics at the very same time took a totally different turn, toward structure and syntax, one that continues to reverberate in language teaching and learning much beyond its initial intellectual appeal.

Of significance for advanced L2 learning are the consequences he identifies for such a reorientation: (1) freeing the grammar from the constraints of structure in the sense that it would not define the lexicogrammatical space but could be expressed 'as a function of its relationships to other features: its line-up in a system, and the interdependency between that system and others'; (2) doing away with a distinction between describing a feature and relating it to other features: 'describing anything consists precisely in relating it to everything else'; (3) modelling language as a resource, not as an inventory; (4) motivating a probabilistic modelling of grammar; and (5) shaping grammar into a lexicogrammar, which would do away with a 'bricks-&-mortar model of a "lexicon" of words stuck together by grammatical cement' (all quotes are from Halliday, 1996: 21). All these concerns continue to reverberate; the direct relevance of such considerations for advanced learning is as obvious as it is profound.

A third way to express the semiotic or meaning orientation of language is through the notion of construal. Indeed, for Halliday, understanding construal is another way of understanding language as a semiotic, meaning-making system (see Halliday 1996, point 5). He identifies three distinct meanings: where language 'constitutes' human experience, the function of grammar is to *construe;* but where language constitutes social processes and social order, the function of grammar is to *enact*, that is, 'the grammar brings about the processes, and the order, through meaning' (*ibid.*: 7). In that case it 'constitutes' in yet a third way, namely by

> creating a parallel universe of its own, a phenomenal realm that is itself made out of meaning. This enables the semiotic process to unfold, through time, in cahoots with material processes, each providing the environment for the other. To put this in other terms, the grammar enables the flow of information to coincide with, and interact with the flow of events. (*ibid.*: 7)

We have here some of the most basic concepts SFL makes available for thinking about language and languaging: language enables us to make meaning of our experiences in the world – its ideational or reflective function; to enact interpersonal relationships – its interpersonal or active function; and to create parallel imagined worlds in texts in a multidimensional space that always involves both the previous metafunctions (cf. Halliday, 1993, point 16).

So, what might any of this mean for 'thinking advancedness' in new ways? I suggest as a first global answer that attending to learners' high level of awareness of the meaning potential that inheres in grammar might, in a deep sense, be the most important awareness any and all instruction can impart, an awareness that has cognitive and affective consequences that would have far-reaching consequences for the instructed L2 environment. In light of the previous comments this is not a minor matter, nor a learning outcome that a single teacher through her stellar teaching might achieve, nor a learning outcome that can be had by concentrating on it at the advanced level, even though we have identified it as being critical at that level. As I have stated for graduate teacher education with a strong content/ meaning orientation, this effort, too, is likely 'to take a village' (Byrnes 2001).

At a less lofty level, the following observations might be made. These learners can be characterized as having a need to mean in appropriate ways in diverse academic, institutional, professional and, generally, public settings, where language use construes the nature of the setting, experientially and interpersonally. Because they will do so by linguistic means being deployed at the textual level, a high level of awareness of the nature of those grammatical resources is crucial. But this is not the kind of 'attention' or 'awareness-' raising that currently drives interactionist research: instead, it is attentiveness to the consequences of languaging choices being made because they 'mean'.

Meaning with lexicogrammar: a continuum

There is little debate about the fact that advanced learners need to expand their vocabulary resources. How that is dealt with, however, is shaped by theoretical understandings of the nature of lexicon in relation to other components of the system.

As already indicated, Halliday's 'grammar' is really 'lexicogrammar' and it is that for reasons motivated by the meaning-making capacity of language. Both lexicon and grammar occupy one stratum of language and do so along a continuum: on the grammar end we find the more open-ended *meaning potential*; on the lexical side we are closer to already *realized meaning*, although it, too, awaits full realization and specification in the utterance itself. How many times have learners been stymied and frustrated because one and the same word has different meanings in what we call 'different contexts'! Privileging grammar, Hasan (1996a) refers to lexicon as a most delicate grammar, in short, a grammarian's dream. Similarly, and beyond SFL, the study of grammaticization, in turn, begins with lexical items and traces their gradual taking on the garment of more general grammatical features.

A powerful instance of the systemic nature of linguistic realization is the construct of grammatical metaphor, a construct that seemingly addresses central issues regarding the phenomenon of advancedness in any language, thus also L2 advancedness. As Halliday indicates, on the semiotic side, it recognizes the shift from the congruent semiosis that characterizes early and mostly spoken language, both ontogenetically and phylogenetically, to the synoptic semiosis that characterizes literate and written language. Where processes are originally expressed dynamically in terms of transitivity patterns – *the country needs immigrants* – they come to be expressed metaphorically, primarily through nominalizations, in written language – *the country's need for immigrants* – thereby transforming them into objects in ideational space that are subject to further meaning-oriented transformations in the clausal space – *the country's increasing/troublesome/unexpected need for immigrants.* As a semiotic capacity it comes relatively late in language acquisition in the development of the adolescent (1993, point 20). At the same time grammatical metaphor is an issue along the continuum of lexical-grammatical space; and finally, it belongs as well in the textual environment, inasmuch as nominalizations become a central tool for establishing logically coherent arguments (see Ryshina-Pankova, this volume).

An additional aspect that deserves mention for advanced learning is the relational contingency between grammar and lexicon. Thus, a simple but profoundly important shift that occurs for one and the same lexical item is that from predicative use of adjectives in a series of clauses to their attributive use in pronominal position. While as a lexical item it is the same it may or may not have different grammatical features (e.g., for German, a complex inflectional apparatus kicks in); but, more important, that lexical item in a different functional context now resides in a different meaning context, that of a clause with new meaning potentialities: We just bought a condominium; unfortunately, it was *completely overpriced.* → No mortgage company is going to be willing to finance a *completely overpriced* condominium.

What might advanced learners gain from such an understanding of language and, more specifically, of its lexicogrammar along a continuum of meaning that has textual consequences? At a first and simple, though certainly not irrelevant, level, it would provide an alternative to the generally negatively experienced demands of rule adherence. This tends to be adherence to morphological and syntactic rules, to things learners 'should know', framed as the demand for 'accurate and automatized control', therefore addressed through practice by means of decontextualized 'drills'. And yet 'mastery' continues to elude them – even after an intermediate level grammar review course that may be followed by an advanced grammar course.

It is tempting to ask whether the kind of 'attention' to meaning that is here intended – an attentiveness that is located at the meaning-form interface of lexicogrammar – would be able to link both meaning and form in a way that advances both, even if that approach were practicable for only some phenomena. FonF advocates would probably assert that is exactly the hallmark of their approach (e.g., Doughty and Williams 1998). However, insistence on a *prior* meaning focus, before formal features are to be brought to the learners' attention, reveals a stance that, deep down, takes our thinking to be non-linguistic, the very point that both SFL and SCT vigorously deny (but see Samuda's ingenious circumvention of that conceptual and pedagogical problem spot, 2001).

In fact, probing a bit more deeply into the consequences of extant practices, it would not be unreasonable to assume a ceiling effect for L2 acquisition within a pedagogical environment that is based on the prevailing and dichotomous understanding of the nature of language. In that case, what instructed learners can and cannot learn, therefore, what levels of L2 ability they can or cannot attain, would at the very least have to be considered in another light – neither the strongly innatist critical period studies nor the overly simplistic cause-and-effect searches for *the* best methodology or, more recently and less idealistically, *best teaching practices.* Simply put, we do not know what learners might be able to learn in an instructional environment that would foreground meaning *within a theory of language that has articulated meaning–form relationships rather than just asserting them.*

Similarly, it strikes me as not unreasonable to consider how a resource orientation for lexicogrammar might become the basis for a kind of intellectual engagement of L2 learners that up to now has not been the descriptor that readily comes to mind for L2 instruction. Whether we call it intellectual, cognitive or meaning-oriented engagement, a 'grammar as resource' stance might significantly reshape the activity of language learning itself, most especially at the advanced level. The motivation literature proposes that 'motivation' translates into increased access to 'input', which results in increased 'interaction', which leads to 'language acquisition'. Within the instructed environment, the desired sequence has been devilishly difficult to prove despite a huge research effort, as Ellis' recent discussion of task-based teaching indicates. He summarizes that 'there is no clear evidence as yet that any of these implementation variables [of tasks] impact on language acquisition' (2003: 100). That fundamental problem aside, with regard to consequences arising from the learner's engagement, rather than the teacher's manipulation of tasks, the stated assumptions depend on *favourable external conditions.* Only then could motivation appreciably affect the nature and quality of the ambient purposes for 'input' purposes.

Instead, language learning in instructed settings, particularly advanced learning, may have to imagine continued language development in terms of expanding learners' *internal meaning-making ability and capacity.* It could do so by creating a learning environment that in its very practices creates in learners high levels of awareness about the 'meaningfulness of grammar', a capacity on their part that recognizes the ambient language of the classroom as much more than 'input'. As both Caffarel and Teruya (this volume) show, learning 'about language' in this fashion might be a particularly promising avenue in instructed advanced learning for learning language itself.

Making meaning: the power of texts in contexts

The third and final excursion into the potential of a refiguration of advanced language learning from the perspective of SFL and SCT is at the level of the text. Once more, there is likely to be little disagreement about the claim that this is what advancedness is all about. In fact, in contrast with the impoverished understanding of grammar and lexicon in SLA thinking, there has been a lively exchange on these matters for quite some time, whether under the rubric of rhetoric or contrastive rhetoric, or under the more skills-oriented notion of enhancing writing, or, more recently from the perspective of literacy and genre in a professional context (see particularly Johns 2002 and the groundbreaking work by Swales 1990 and 2004). It is fair to say that much of that work offers important insights for the advanced L2 context as well.

But, just as SFL offers unique perspectives for our understanding of grammar or lexicon that capture important aspects of advanced learning so, too, for textuality. Once more, a miniscule glimpse of that potential will have to suffice. When I stated at the beginning of this paper that SFL probably was in a category of its own among theories of language in terms of its interest in educational issues, the same could be said for its interest in texts and contexts, once again an interest that is foundational rather than subsequently grafted on (see e.g., Hasan 1995, 1996b, d, e). Thus Halliday's central exposition of the theory (1985/1994) states unequivocally that his aim has been to 'construct a grammar for purposes of text analysis: one that would make it possible to say sensible and useful things about any text, spoken or written, in modern English' (1985/1994: xv). That usefulness would begin at the level of understanding the text in terms of a linguistic analysis 'to show how, and why, the text means what it does. In the process, there are likely to be revealed multiple meanings, alternatives, ambiguities, metaphors, and so on' (*ibid.*). At the next higher level one would aim for an evaluation of the text, a stance that would determine its effectiveness or not. As Halliday emphasizes, that kind of analysis is considerably more complex inasmuch as it requires the inclusion of contextual features, what SFL theory, drawing on Malinowski's earlier distinctions, refers to as the context of situation and the context of culture. Critically, this is not some sort of fuzzy claim for 'cultural embeddedness', all too frequently the placeholder for a sophisticated analysis of 'context': Halliday unmasks it as little more than 'running commentary' (1985/1994: xvi): 'without "a theory of wordings" – that is, a grammar – there is no way of making explicit one's interpretation of the meaning of a text' (1985/1994: xvii). That stipulation also specifies the kind of grammar that is needed: a discourse grammar that is both functional and semantic in its orientation in order to show up how grammatical categories and choices result in semantic patterns.

While that may be more than many of us would have bargained for, it seems that, for the sake of rigorous analysis of texts, both a careful analysis of 'context' and a way to relate the textual organization to that outer context are necessary. SFL provides a highly developed theoretical and practical system for accomplishing both. To gain a first sense, I refer the reader to Matthiessen's careful discussion of context, particularly the theoretical status of the context of situation as construed, as contrasted with the physical realities that may attend to a setting: it relates texts to the social processes within which it is located. In turn that context of situation resides within a larger context of culture that enables the linguistic construal of situations in the first place (see also the lucid treatment by Hasan 1995, 1996e, 1999b).

Perhaps, little more can and should be said here, except for the following characterization of what advanced learning would then be all about because of its focus on texts:

> In any situation involving language and learning, you have to be able to move in both directions: to use the situation to construe the text, as Malinowski did, but also to use the text as a means to construe the situation. The situation, in other words, may not be something that is 'given'; it may have to be construed out of the text . . . The term that we usually use for this relationship, coming from European functional linguistics, is *realization:* the situation is 'realized' in the text. Similarly the culture is 'realized' in the linguistic system. This does not mean that the one somehow *causes* the other. The relation is not one of cause. It is a semiotic relationship: one that arises between pairs of information systems, interlocking systems of meaning . . . Thus the culture is construed by systems of language choice; the situation is construed by patterns of language use. (Halliday 1999b: 14–15, original emphases)

I can think of few better ways to describe the challenge, opportunity and intellectual excitement for advanced learners engaging with texts in such a fashion.

Grammar, lexicon and texts are, of course, vast categories and I have only presented their possible reconceptualization for advanced learning in the sketchiest of ways. As a way of hinting at the enormous potential for exploration beyond these considerations, I pose the following questions in no particular order and without further commentary.

- Can and should an explicitly meaning-oriented approach characterize an entire language programme, *ab initio* to upper levels of L2 ability, or is it more appropriate after lexicogrammatical resources have reached a certain breadth, depth and confident accessibility for the learner? Is it possible to describe that stage more closely and on what basis?
- Assuming that language is a tool for meaning-making, how does advanced L2 teaching address the tension between 'gaining the freedom to create', to establish new identities, and to self-regulate through linguistics means (Dunn and Lantolf, 1998) and learning how to mean generically and how might this be specified within curricular levels and across them?
- What is the relation between an SFL approach and a cognitive semantic approach, (e.g., Langacker, Slobin)?
- Given that information structuring that is based on L1 patterns of grammaticization seems to linger into very advanced L2 abilities and seems to affect ultimate attainment, can such attainment be facilitated in substance or in terms of earlier acquisition through conscious teaching of some of the interrelations at the level of lexicogrammar, looking from above, looking from below and looking from within, as Matthiessen describes it?
- What degree of specificity should characterize an L2 program, for instance a four-year collegiate program, for such an approach to be translated into educational practice?
- Is it necessary/advantageous for literate adult learners to follow the same sequence of semiosis, from more congruent forms to more synoptic forms? Related to that, assuming that the shift in semiosis, from congruent to synoptic, characterizes most literate languages studied in educational

programs, what kinds of allowances should or must be made for the acquisition of different target languages as they are paired with different source languages?

- Is that shift realized in similar ways across languages, e.g., primarily through nominalizations and downward rank-shifting interclausal relations into intraclausal relations? Are there certain topologies that can be observed?

A reflective coda

I return to the central question of this introduction and the entire volume: is a link between SFL, SCT and advanced L2 learning 'meaning-ful' at this point? I answer that question with reference to Vygotsky's notion of the zone of proximal development (ZPD). In doing so I follow Hasan's example (2005b), who invoked the construct as a way of deepening the insights to be gained by linking Vygotsky's sociogenetic, tool-mediated theory of mind, Halliday's sociological linguistics and Bernstein's analysis of different forms of semiotic mediation in one and the same social community. Similarly, I will invoke it at the confluence of SFL, SCT and advanced L2 learning, based on the strong belief that it can provide the kind of mediational environment and also the conceptual tools with which the field might expand its horizons in order to (re-)gain control over an intellectual and practice-oriented environment, particularly in SLA research, that its current approaches seem unable to address.

Does it offer stimulating and viable ways for following up on the recommendations in Firth and Wagner (1997) when they sought an enlarged social and contextual understanding of language, language acquisition research and pedagogical practice? Does it also offer a way to respond to Ortega's (2005a and b) recent call, close to a decade after Firth and Wagner, that the SLA field needs to be broadened, not least because 'instructed SLA research is (or should be) research that inhabits, and is reflective of, a diversity of educational contexts and that is inspired by the goal to improve learning and teaching in the full spectrum of educational contexts where L2s matter' (2005b: 318)? Finally, does it offer new ideas for a seemingly deadlocked discussion among higher education professionals who are ambivalent about the appropriateness of the prevailing paradigm of communicative competence, at least as practised, with regard to desirable learning goals and who are even more concerned about its sufficiency as an intellectually viable goal for higher education foreign language programs (Byrnes 2006)? I leave it to the readers of this volume to answer those questions.

Notes

1 I am aware that Lantolf, in particular, has chosen advanced language learning as the focus of the federally funded national language resource centre he directs, the Center for Advanced Language Proficiency, Education, and Research, CALPER.

However, the distinction I am here making is between what I perceive to be the Center's interest in diverse areas of advancedness, e.g., most recently the highly charged area of assessment, where these are often seen in light of Vygotskian sociocultural theory (see, e.g., much of the research that Lantolf himself has supervised and that is referenced in many of his publications) and an approach that reveals an overarching sense of the nature of advancedness itself and how it differs from other acquisitional levels. Of course, a counter-argument would be that highlighting such a difference is only an artifact of current instructional practices. Even so, I believe that advancedness will continue to have special distinguishing foci and characteristics that are worthy of careful consideration.

2 Thus, among many other examples that could be cited, Fauconnier's relatively accessible treatment of how we go about constructing meaning in thought and language, *Mappings in thought and language*, has no mention of Halliday's work, even as he decries the habit of 'modern work' in linguistics and philosophy of 'attempting to study the grammatical or meaning structure of expressions independently of their function in building up discourse, and independently of their use in reasoning and communication' (1997: 5). Similarly, Langacker, in one of his most recent treatments of central issues in a functionally and semantically oriented cognitive linguistics and one that explicitly invited consideration of advanced L2 learning, treated matters primarily in terms of '**mental constructions** that intervene between the situations we describe and the form and meaning of the expressions employed', adding that 'cognition is also **contextually embedded**' (2006: 17, original emphases).

3 As Babich (2003) points out, the word 'paradigm' is by no means innocent. Used by Kuhn in his enormously influential *The Structure of Scientific Revolutions* (1962), it constitutes a deliberate terminological choice by Kuhn at that time when the Cold War was at its peak. In particular, it avoided the problematic though surely obvious 'thought collective' as a translation for the German word *Denkkollektiv*, a term that plays a central role in the essay which, as Kuhn acknowledges in the preface, 'anticipates many of my own ideas' (1962: 7). Published in German in 1935 by the Polish microbiologist Ludwik Fleck, that essay carefully traces over several centuries the evolution of syphylis as a socially and culturally constructed fact, even in the environment of the natural sciences. Babich notes that such a challenge to Western individualistic notions of how we go about knowing called forth Kuhn's conceptual way of expressing matters, thereby giving us the aforementioned 'paradigm' – and also 'structures' of 'revolutions'. Much like Dunn and Lantolf (1998) had declared the ZPD as incommensurable with Krashen's notion of i+1, so Babich observes an incommensurability between the 'thought collective' and a 'paradigm'.

4 For an interesting theoretical discussion of process and product issues in twentieth-century linguistics that draws particularly on Whorf in relation to SFL theory, see Hasan 1996c from whom my own reflections, as their title readily indicates, have benefited greatly; also her trenchant review of Bourdieu's notion of literacy, occasioned by Bourdieu's *Language and Symbolic Power* (1999a).

5 Some brief comments on the level of 'mechanics' for this section are in order: I suspect readers will already have noticed various forms of intertextual borrowings. Thus, the title of the volume itself arose from my early reading of Wells' comparison (1994) of Halliday's and Vygotsky's central concerns. Similarly the title I have chosen for my introductory comments is an expansion of Hasan's reflection on language as a resource (1996c). And there are likely to be many more such

borrowings, blendings, covert and overt dialogues and imitations that I may no longer even be aware of! Finally, a densely argued framework like SFL makes it difficult to separate what must be incorporated as a direct quote from what can appropriately be included, with attribution, as a paraphrase; in my decisions I hope to have come down 'where I ought to be'.

References

Babich, B. (2003) 'Kuhn's paradigm as a parable for the Cold War: incommensurability and its discontents from Fuller's Tale of Harvard to Fleck's Unsung Lvov'. *Social Epistemology*, 17, 99–109.

Behrens, B. (2006) 'Language-based processing in advanced L2 production and translation: an exploratory', in H. Byrnes, H. Weger-Guntharp and K. A. Sprang (eds), pp. 74–86.

Block, D. (2003) *The Social Turn in Second Language Acquisition*. Washington, DC: Georgetown University Press.

Brown, J. D. (1995) *The Elements of Language Curriculum. A Systematic Approach to Program Development*. Boston: Heinle & Heinle.

Byrnes, H. (1998) 'Constructing curricula in collegiate foreign language departments', in H. Byrnes (ed.), *Learning Foreign and Second Languages: Perspectives in Research and Scholarship*. New York: MLA, pp. 262–95.

Byrnes, H. (2001) 'Reconsidering graduate students' education as teachers: it takes a department!' *Modern Language Journal*, 85, 512–30.

Byrnes, H. (2002a) 'The cultural turn in foreign language departments: challenge and opportunity'. *Profession 2002*, 114–29.

Byrnes, H. (2002b) 'Toward academic-level foreign language abilities: reconsidering foundational assumptions, expanding pedagogical options', in B. L. Leaver and B. Shekhtman (eds), pp. 34–58.

Byrnes, H. (2004) 'Advanced L2 literacy: beyond option or privilege'. *ADFL Bulletin*, 36(1), 52–60.

Byrnes, H. (2005a) 'Content-based foreign language instruction', in C. Sanz (ed.), *Mind and Context in Adult Second Language Acquisition: Methods, Theory, and Practice*. Washington, DC: Georgetown University Press, pp. 282–302.

Byrnes, H. (2005b) 'Literacy as a framework for advanced language acquisition'. *ADFL Bulletin*, 37(1), 11–15.

Byrnes, H. (2006) 'Perspectives: Interrogating communicative competence as a framework for collegiate foreign language study'. *Modern Language Journal*, 90, 244–66.

Byrnes, H. (2007) 'Language acquisition and language learning', in D. G. Nicholls (ed.), *Introduction to Scholarship in Modern Languages and Literatures* (3rd edn), New York: MLA.

Byrnes, H. and Kord, S. (2002) 'Developing literacy and literary competence: Challenges for foreign language departments', in V. M. Scott and H. Tucker (eds), *SLA and the Literature Classroom: Fostering Dialogues*. Boston: Heinle & Heinle, pp. 31–69.

Byrnes, H. and Maxim, H. H. (eds) (2004) *Advanced Foreign Language Learning: A Challenge to College Programs*. Boston, MA: Heinle Thomson.

Byrnes, H. and Sprang, K. A. (2004) 'Fostering advanced L2 literacy: a genre-based, cognitive approach', in H. Byrnes and H. H. Maxim (eds), pp. 47–85.

Byrnes, H., Crane, C., Maxim, H. H. and Sprang, K. A. (2006) 'Taking text to task:

Issues and choices in curriculum construction', *International Journal of Applied Linguistics*, 19, 2, 1–25.
Byrnes, H., Weger-Guntharp, H. and Sprang, K. A. (eds) (2006) *Educating for Advanced Foreign Language Capacities: Constructs, Curriculum, Instruction, Assessment.* Washington, DC: Georgetown University Press.
Carroll, M. and Lambert, M. (2006) 'Reorganizing principles of information structure in advanced L2s: French and German learners of English', in H. Byrnes, H. Weger-Guntharp and K. A. Sprang (eds), pp. 54–73.
Christie, F. and Martin, J. R. (eds) (1997) *Genre and Institutions: Social Processes in the Workplace and School.* London: Cassell.
Christie, F. and Unsworth, L. (2005) 'Developing dimensions of an educational linguistics', in R. Hasan, C. Matthiessen and J. J. Webster (eds), pp. 217–50.
Cloran, C., Butt, D. and Williams, G. (eds) (1996) *Ways of Saying: Ways of Meaning. Selected Papers of Ruqaiya Hasan.* London: Cassell.
Colombi, M. C. and Schleppegrell, M. J. (2002) 'Theory and practice in the development of advanced literacy', in M. J. Schleppegrell and M. C. Colombi (eds), pp. 1–19.
Doughty, C. and Williams, J. (eds) (1998) *Focus on Form in Classroom Second Language Acquisition.* Cambridge: Cambridge University Press.
Dunn, W. E. and Lantolf, J. P. (1998) 'Vygotsky's zone of proximal development and Krashen's i + 1: incommensurable construct; incommensurable theories'. *Language Learning*, 48, 411–42.
Ellis, R. (2003) *Task-based Language Learning and Teaching.* Oxford: Oxford University Press.
Fauconnier, G. (1997) *Mappings in Thought and Language.* New York: Cambridge University Press.
Firth, A. and Wagner, J. (1997) 'On discourse, communication, and (some) fundamental concepts in SLA research'. *Modern Language Journal*, 81, 285–312.
Fleck, L. (1979) T. J. Trenn and R. K. Merton (eds), *Genesis and Development of a Scientific Fact.* Chicago: University of Chicago Press.
Freed, B., So, S. and Lazar, N. A. (2003) 'Language learning abroad: how do gains in written fluency compare with gains in oral fluency in French as a second language?'. *ADFL Bulletin*, 34(3), 34–40.
Hall, J. K. (2002) *Methods for Teaching Foreign Languages: Creating a Community of Learners in the Classroom.* Upper Saddle River, NJ: Prentice Hall.
Halliday, M. A. K. (1973) *Explorations in the Functions of Language.* London: Edward Arnold.
Halliday, M. A. K. (1985/1994) *An Introduction to Functional Grammar.* London: Edward Arnold.
Halliday, M. A. K. (1993) 'Towards a language-based theory of learning'. *Linguistics and Education*, 5, 93–116.
Halliday, M. A. K. (1996) 'On grammar and grammatics', in R. Hasan, C. Cloran and D. G. Butt (eds), *Functional Descriptions: Theory in Practice.* Amsterdam/Philadelphia: John Benjamins, pp. 1–38.
Halliday, M. A. K. (1999a) 'Grammar and the construction of educational knowledge', in R. Berry, B. Asker, K. Hyland and M. Lam (eds), *Language Analysis, Description and Pedagogy.* Hong Kong: Language Centre, The Hong Kong University of Science & Technology and Department of English, Lingnan University, pp. 70–87.
Halliday, M. A. K. (1999b) 'The notion of "context" in language education', in M. Ghadessy (ed.), *Text and Context in Functional Linguistics.* Amsterdam/Philadelphia: John Benjamins, pp. 1–24.

Halliday, M. A. K. (2006) 'Systemic theory', in K. Brown (ed.), *Encyclopedia of Languages and Linguistics*. Amsterdam: Elsevier, pp. 443–8.

Halliday, M. A. K. and Matthiessen, C. M. I. M. (2004) *An Introduction to Functional Grammar*. London: Edward Arnold.

Hasan, R. (1995) 'The conception of context in text', in P. H. Fries and M. Gregory (eds), *Discourse in Society: Systemic Functional Perspectives: Meaning and Choice in Language: Studies for Michael Halliday*. Norwood, NJ: Ablex, pp. 183–283.

Hasan, R. (1996a) 'The grammarian's dream: lexis as a most delicate grammar', in C. Cloran, D. Butt and G. Williams (eds), pp. 73–103.

Hasan, R. (1996b) 'Literacy, everyday talk and society', in R. Hasan and G. Williams (eds), *Literacy in Society*. London: Longman, pp. 377–424.

Hasan, R. (1996c) 'Speech genre, semiotic mediation and the development of higher mental functions', in C. Cloran, D. Butt and G. Williams (eds), pp. 152–90.

Hasan, R. (1996d) 'What kind of resource is language?', in C. Cloran, D. Butt and G. Williams (eds), pp. 13–36.

Hasan, R. (1996e) 'What's going on: a dynamic view of context in language', in C. Cloran, D. Butt and G. Williams (eds), pp. 37–50.

Hasan, R. (1999a) 'The disempowerment game: Bourdieu and language in literacy'. *Linguistics and Education*, 10, 25–87.

Hasan, R. (1999b) 'Speaking with reference to context', in M. Ghadessy (ed.), *Text and Context in Functional Linguistics*. Amsterdam/Philadelphia: John Benjamins, pp. 219–318.

Hasan, R. (2005a) 'On the social conditions for semiotic mediation: the genesis of mind in society', in J. J. Webster (ed.), pp. 106–29.

Hasan, R. (2005b) 'Semiotic mediation and three exotropic theories: Vygotsky, Halliday and Bernstein', in J. J. Webster (ed.), pp. 130–56.

Hasan, R., Matthiessen, C. and Webster, J. J. (eds) (2005) *Continuing Discourse on Language: A Functional Perspective* (Vol. 1). London: Equinox.

Hasan, R. and Williams, G. (eds) (1996) *Literacy in Society*. London: Longman.

Johns, A. M. (2002) *Genre in the Classroom: Multiple Perspectives*. Mahwah, NJ: Lawrence Erlbaum.

Kuhn, T. S. (1996) *The Structure of Scientific Resolutions* (3rd edn). Chicago: University of Chicago Press.

Langacker, R. W. (1998) 'Conceptualization, symbolization, and grammar', in M. Tomasello (ed.), *The New Psychology of Language: Cognitive and Functional Approaches to Language Structure* (Vol. 1). Mahwah, NJ: Lawrence Erlbaum, pp. 1–39.

Langacker, R. W. (2006) 'The conceptual basis of grammatical structure', in H. Byrnes, H. Weger-Guntharp and K. A. Sprang (eds), pp. 17–39.

Lantolf, J. P. (2000a) 'Second language learning as a mediated process'. *Language Teaching*, 33, 79–96.

Lantolf, J. P. (2000b) *Sociocultural Theory and Second Language Learning*. Oxford: Oxford University Press.

Lantolf, J. P. (2006) 'Sociocultural theory and L2: state of the art'. *Studies in Second Language Acquisition*, 28, 67–109.

Lantolf, J. P. and Appel, G. (eds) (1994) *Vygotskian Approaches to Second Language Research*. Norwood, NJ: Ablex.

Leaver, B. L. and Shekhtman, B. (2002) *Developing Professional-level Language Proficiency*. Cambridge: Cambridge University Press.

Long, M. H. (1990) 'The least a second language acquisition theory needs to explain'. *TESOL Quarterly*, 24, 649–66.

Lucy, J. A. and Wertsch, J. V. (1987) 'Vygotsky and Whorf: a comparative analysis', in M. Hickmann (ed.), *Social and Functional Approaches to Language and Thought*. Orlando, FL: Academic Press, pp. 67–85.

Malinowski, B. (1935) *Coral Gardens and their Magic*. Vol. 2. London: Allen & Unwin.

Martin, J. R. (1997) 'Analysing genre: functional parameters', in F. Christie and J. R. Martin (eds), pp. 3–39.

Martin, J. R. (1999) 'Mentoring semogenesis: "genre-based" literacy pedagogy', in F. Christie (ed.), *Pedagogy and the Shaping of Consciousness: Linguistic and Social Processes*. London: Cassell, pp. 123–55.

Martin, J. R. (2000) 'Design and practice: enacting functional linguistics'. *Annual Review of Applied Linguistics*, 20, 116–26.

Mohan, B. A. (1986) *Language and Content*. Reading, MA: Addison-Wesley.

Ortega, L. (2005a) 'For what and for whom is our research? The ethical as transformative lens in instructed SLA'. *Modern Language Journal*, 89, 427–43.

Ortega, L. (2005b) 'Methodology, epistemology, and ethics in instructed SLA research: an introduction'. *Modern Language Journal*, 89, 317–27.

Ortega, L. and Byrnes, H. (eds) (2007) *The Longitudinal Study of Advanced L2 Capacities*. Mahwah, NJ: Lawrence Erlbaum.

Rothery, J. (1996) 'Making changes: developing an educational linguistics', in R. Hasan and G. Williams (eds), pp. 86–123.

Samuda, V. (2001) 'Guiding relationships between form and meaning during task performance: the role of the teacher', in M. Bygate, P. Skehan and M. Swain (eds), *Researching Pedagogic Tasks: Second Language Learning, Teaching and Testing*. Harlow, England: Pearson Education Limited, pp. 119–40.

Schleppegrell, M. J. (2004) *The Language of Schooling: A Functional Linguistics Perspective*. Mahwah, NJ: Lawrence Erlbaum.

Schleppegrell, M. J. and Colombi, M. C. (2002) *Developing Advanced Literacy in First and Second Languages: Meaning with Power*. Mahwah, NJ: Lawrence Erlbaum.

Slobin, D. I. (1996) 'From "thought and language" to "thinking for speaking" ', in J. J. Gumperz and S. C. Levinson (eds), *Rethinking Linguistic Relativity*. Cambridge: Cambridge University Press, pp. 70–96.

Stutterheim, C. von and Carroll, M. (2006) 'The impact of grammatical temporal categories on ultimate attainment in L2 learning', in H. Byrnes, H. Weger-Guntharp and K. A. Sprang (eds), pp. 40–53.

Swaffar, J. (2004) 'A template for advanced learner tasks: Staging genre reading and cultural literacy through the précis', in H. Byrnes and H. H. Maxim (eds), pp. 19–45.

Swales, J. M. (1990) *Genre Analysis: English in Academic and Research Settings*. Cambridge: Cambridge University Press.

Swales, J. M. (2004) *Research Genres: Explorations and Applications*. New York: Cambridge University Press.

Vygotsky, L. (1978) *Mind in Society. The Development of Higher Psychological Processes*. M. Cole, V. John-Steiner, S. Scribner and E. Souberman (eds). Cambridge, MA: Harvard University Press.

Vygotsky, L. (1986) *Thought and Language*. Cambridge, MA: MIT Press.

Webster, J. J. (ed.) (2005) *Language, Society and Consciousness. The Collected Works of Ruqaiya Hasan*. Vol. 1. London: Equinox.

Wells, G. (1994) 'The complementary contribution of Halliday and Vygotsky to a "language-based theory of learning" '. *Linguistics and Education*, 6, 41–90.

Wertsch, J. V. (1985) *Vygotsky and the Social Formation of Mind.* Cambridge, MA: Harvard University Press.

Wertsch, J. V. (1990) 'Dialogue and dialogism in a socio-cultural approach to mind', in I. Markova and K. Foppa (eds), *The Dynamics of Dialogue.* New York: Harvester Wheatsheaf, pp. 62–82.

Wertsch, J. V. (1991) *Voices of the Mind: A Sociocultural Approach to Mediated Action.* Cambridge, MA: Harvard University Press.

Wertsch, J. V. (1998) *Mind as Action.* New York: Oxford University Press.

Wertsch, J. V. (2000) 'Vygotsky's two minds on the nature of meaning', in C. D. Lee and P. Smagorinsky (eds), *Vygotskian Perspectives on Literacy Research: Constructing Meaning through Collaborative Inquiry.* Cambridge: Cambridge University Press, pp. 19–30.

Williams, G. (2005) 'Grammatics in schools', in R. Hasan, C. Matthiessen and J. J. Webster (eds), pp. 281–310.

Part I

Theoretical Considerations in Advanced Instructed Learning

1 Educating for advanced foreign language capacities: exploring the meaning-making resources of languages systemic-functionally

Christian M. I. M. Matthiessen

1 Conception of language and language education

The approach that we adopt to any of the phases of learning a second or foreign language will depend on a range of factors – including the context of learning/teaching, the age of the learners, the range of learning styles within a given group of learners, and our conception of language.

The last factor is critical, but it sometimes slips into the background in discussions of language education. It is critical because it will determine what is brought into focus in the processes of language teaching and learning, how these processes are staged in the curriculum, and what kinds of material are used and developed to support them. For example, many aspects of the initiatives in language teaching in the United States during and after World War II were determined by the nature of the American structuralist linguistics of the day. It is also this conception that will determine how we shift our focus in the move from novice and intermediate learners towards advanced learners. As research into learning progresses, it will be possible to relate the conception of language to a language-based theory of learning (cf. Halliday 1993; Painter 1999).

The point is: our approach to the teaching of any phenomenon depends critically on our conception of this phenomenon. Unless we can base language teaching and learning on a richly revealing comprehensive account of what kind of phenomenon language is, we are not in a position to answer the many questions that arise in educational contexts. Similarly, the value and success of any approach designed to support second or foreign language teaching, such as contrastive analysis or error analysis, will depend critically on the conception of language that informs it. For example, Lado's (1957) pioneering approach to contrastive analysis was based on an American structuralist conception of language as rule-governed patterns, and it is very different from a systemic functional approach based on a conception of language as a resource for making meaning – an approach that can be illuminated, among others, by Lantolf and Thorne's (2006) research into the mediating role played by the semantic system of the mother tongue in second language development.

Many years ago, B. L. Whorf (1956: 207–8) characterized the commonsense conception of language as the position of 'natural logic', and what he said is quite relevant here, including his characterization of 'natural logic' as conceiving of talking as 'merely an incidental process concerned with communication, not with formulation of ideas'. As shown in Matthiessen (1993b, 1998) and Halliday and Matthiessen (1999: Chapter 14), this view of language is itself based on the unconscious folk theory that is embodied in our everyday grammar, which is a key reason why it is typically assumed rather than problematized and challenged. According to 'natural logic', perception and conceptualization are independent of, and precede, verbalization. If that is so, learning a new language is only a matter of learning a new way of communicating language-independent ideas, and the learning of language and the learning of content can be conceived of as distinct activities.

Whorf's own view was, of course, diametrically opposed to that of 'natural logic'. He gave language a much more central role in the construal of experience – in the formation of ideas, in thinking and in reasoning – a position that is related to his notion of fashions of speaking and frames of consistency (cf. Martin 1988: 246–52). Indeed, since Whorf's time, powerful new evidence has come to light showing the fallacy of the notion that perception and conceptualization are independent of, and precede, verbalization in a given language. For example, investigating the specific conceptual domain of event construal, von Stutterheim and Carroll (2006: 41) present evidence 'for the interrelation between grammaticized means and specific principles of information organization'. Through experiments involving speakers of different languages reporting sequences of events in film clips, they were able to demonstrate that their languages guided their processes of perception and conceptualization from the start (brought out, for example, in the different patterns of eye movements characteristic of speakers of different languages). With regard to second language learning they conclude: 'the central factor impeding the acquisitional process at advanced stages ultimately is grammatical in nature, in that learners have to uncover the role accorded to grammaticized meanings and what their presence, or absence, entails in information organization' (2006: 51).

Similarly, Lantolf (this volume, citing Negueruela *et al.*) reports on how, through an analysis of gestures accompanying language, one can uncover how even advanced second language learners construe events in their second language based on the semantic system of their first language. The effects can be quite subtle and 'errors' can be hard to detect, but closer analysis reveals semiotic mediation of the first language. Once we can identify and interpret such effects on 'semantic style' in a second language through a Vygotskian/Whorfian/Hallidayan perspective, we are in a position to help advanced language learners develop their semantic resources in the language they are learning.

More generally, we now seem to have entered a phase of unprecedented opportunities for moving ahead in very productive ways across a range of activities, including centrally second and foreign language education: insights from Vygotsky and Leontiev (with their activity theory), Bakhtin, Bruner, Wittgenstein, Whorf, Hjelmslev, Malinowski, Firth, Bernstein and Halliday's systemic functional linguistics; and the breakthroughs in a semantically oriented psycholinguistic research referred to above, as well as the new focus on language and the brain by leading neuroscientists such as Deacon, Edelman and Arbib, resonate with one another, contributing complementary strands. These insights guide and are also extended by longitudinal studies of how people learn how to mean – children learning how to mean in their protolanguages and moving into their mother tongues (e.g., Halliday 1975, 2003; Painter 1999), and students gradually learning how to mean in a second/foreign language (see some aspects of longitudinal development in the contributions by Colombi, Crane and Ryshina-Pankova, this volume, and the papers in Ortega and Byrnes 2007). A number of points of contact and cross-fertilization among these strands have already been highlighted, as in Steiner's (1991) integration of Leontiev's theory of action into a systemic functional approach, and Wells' (1999) and Williams' (2005) discussion of Halliday and Vygotsky. This volume is another indication of the extensive potential for further developments. In this paper, I will explore some aspects of the systemic functional conception of language as a resource for making meaning, highlighting, as appropriate, the special case of the advanced learner.

2 Three aspects of learning language

In learning a foreign language (see Section 3), a learner is also learning ***through*** the language and learning ***about*** the language – to put this in terms of M. A. K. Halliday's simple but powerful schema for thinking about language learning (e.g., Halliday 1980; Hasan and Martin 1989; Mohan 1986, 1989; Painter 1999). Accordingly, I will outline the key parameters in terms of which we can reason about advanced language learners in reference to learning a language, learning through a language and learning about a language, treating these as complementary and mutually supportive aspects of learning a language. Indeed, the more advanced learners become, the more these three fundamental aspects of learning a language impinge on one another. Learning a language increasingly becomes a matter of learning through this language in a growing range of quotidian and professional contexts (thus moving closer to the condition of native speakers); and learning a language can increasingly be helped by learning about this language – not only passively, but also actively by investigating it and by developing one's own resources for learning.

Furthermore, as Table 1.1 below indicates, learning through language is intimately linked to the expansion of a learner's registerial repertoire (Section 3.4.2) and can be guided by a context-based typology of texts/registers (Section 3.4.3). This also relates to the important systemic functional work by

Table 1.1 Advanced learners learning language, learning through language and learning about language

learning domain	dimension		advanced learners
learning language	global dimensions	stratification	lower strata falling into place as automated realizations – increasing focus on meaning in context, and on learning to mean in ways different from those in the mother tongue; but lower-stratal patterns of new registers (including, almost certainly, grammatical metaphor), and 'remedial' work, e.g., on prosodic patterns of intonation and rhythm
		instantiation	dialectic of system and text (process); moving up the cline of instantiation to expand 'personalized' meaning potential by learning new registers in new situation types/ institutional domains, very probably more specialized ones; learning lower-frequency parts of the system (terms that are more marked in terms of probability of instantiation); process of instantiation: new kinds of instantiation in addition to speaking/writing and listening/reading, including editing, translating, transcribing
		metafunction	possibly 'remedial' work on textual metafunction (management of the flow of information) and on the appraisal systems of the interpersonal metafunction; learning more of the logical mode of the ideational metafunction in clause complexing as a resource for rhetorical organization (cf. Teruya, this volume); expanding the ideational and interpersonal resources through grammatical metaphor (see Colombi and Ryshina-Pankova, this volume)
	local dimensions	rank	more focus on the higher ranks of the semantic stratum; very possibly 'remedial' work on the higher ranks of phonology (relating to intonation and rhythm)
		axis	learning more marked terms within the overall system; learning lexicogrammatical patterns of intermediate delicacy (in the region between grammar and lexis – 'constructions')
learning through language			learning through texts in the foreign language, in relation to both field (domains of experience) and tenor (social roles and values): as learners become more advanced, such texts can move closer to where they are at in their general learning/professional experience, supplementing such learning through the mother tongue
learning about language			learning more about language (including language as a resource for learning), and the skills to monitor and diagnose one's own text; learning to think grammatically; learning to use tools for investigating language and for producing resources for one's own processes of learning

Mohan and his research group showing how students are helped by learning to 'translate' between language and other semiotic systems (e.g., Mohan 1979, 1986, 1989; Mohan and Zingzi 2002; see also Ryshina-Pankova, this volume).

Learning about language, in turn, is part of becoming a more autonomous learner – quite probably, a lifelong learner of the language. The key principle is to empower the learner, and this includes a range of strategies, both computational and theoretical. Computational tools such as Wu's (2000) SysConc and our database system for developing one's own text archive can play a key role in enabling learners to investigate areas of the language they are learning. Alongside the development of computational tools, the development of new theoretical and descriptive 'tools', such as those of systemic functional linguistics sketched here, is making a major difference in educating for advanced foreign language capacities, as a number of papers in this volume document.

3 Learning language: learning how to mean

3.1 Learning language as learning how to mean

Learning a new language means ***learning how to mean in that language*** – learning the resources for making meaning in context.[1] Learning language is a **multi-dimensional process** – proceeding along a number of intersecting **semiotic dimensions** that define semiotic space (set out in Table 1.1 under the headings of 'global' and 'local' dimensions; see also Figure 1.1). In the course of learning a foreign language, learners are able to expand into this semiotic space, thereby expanding their own personal meaning potentials.

By modelling language in context in comprehensive terms as a multidimensional semiotic space, we create a **map** that can be a key resource in advanced language education (for the metaphor of cartography, cf. Matthiessen 1995) – or rather, a set of complementary maps forming a **semiotic atlas** (Butt, p.c.). Such maps have been indexed in a set of two-dimensional matrices – the function-rank matrix (e.g., Halliday 1976, 1978; Halliday and Matthiessen 2004), the stratification-instantiation matrix (e.g., Halliday 2002), and the function-stratification matrix (cf. Matthiessen *et al.* 2005); an example of a combined function-stratification matrix and function-rank matrix is given in Table 1.2 on p. 40. This cartographic approach makes it possible to ***chart courses for the journeys of learning***. It can guide curriculum and syllabus design, materials development, diagnosis and treatment of errors at the frontier of the learning journey, teaching and learning strategies. Relevant to all the phases of the processes of language education, it can give both teachers and advanced students more control over the teaching–learning process.

I will now discuss the three global dimensions of organization mentioned above – the hierarchy of stratification (Section 3.2), the spectrum of metafunction (Section 3.3) and the cline of instantiation (Section 3.4), and

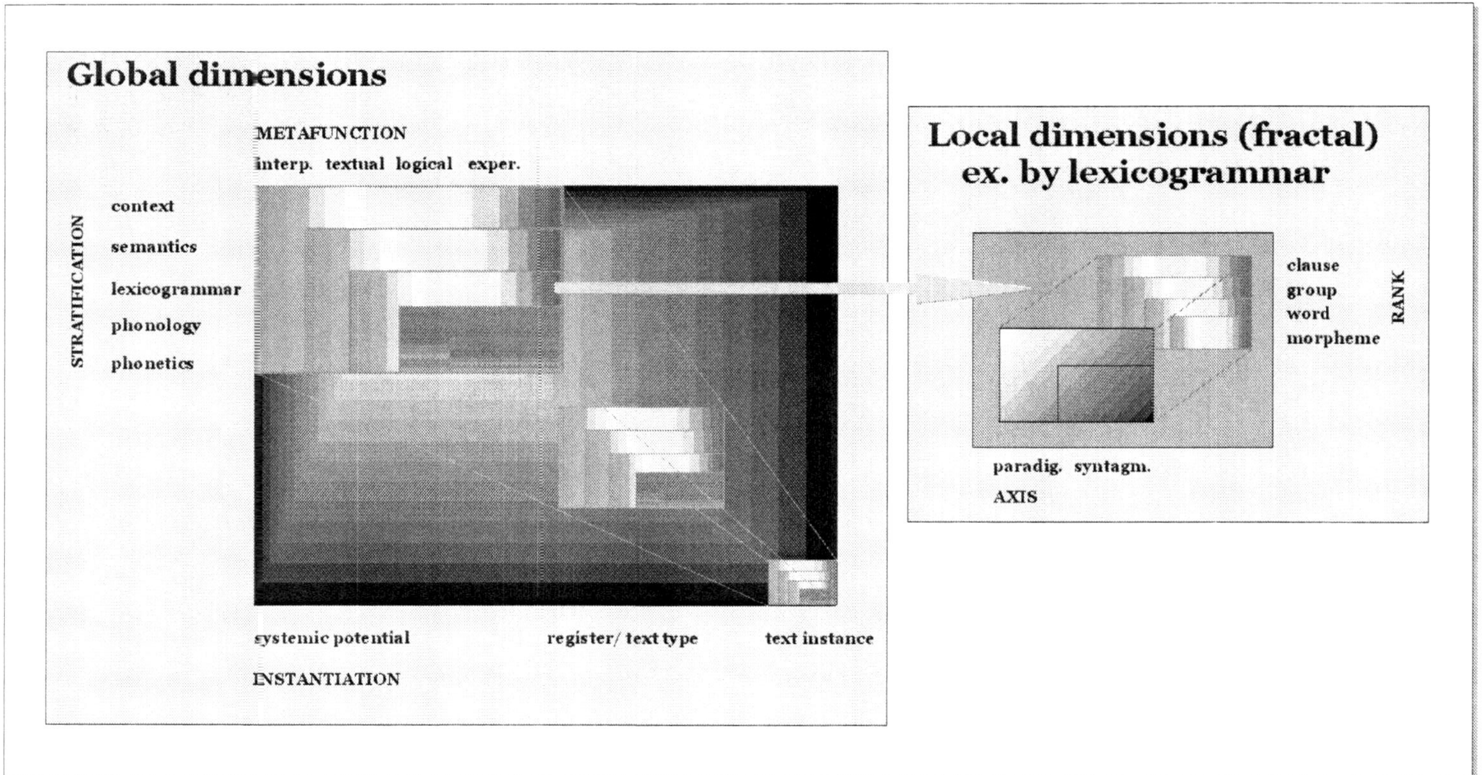

Figure 1.1 The dimensions organizing language in context – global dimensions, and local ones manifested fractally within each stratal subsystem

then turn to dimensions that are local to a given stratal subsystem, such as phonology or lexicogrammar (Section 3.5).

3.2 The hierarchy of stratification: learning how to mean trinocularly

Language is located within context – its semiotic environment; the relationship between the two is a stratal one, which is the same kind of relationship on which the internal organization of language is based (Halliday 1978). Internally, **language** is stratified into two content strata – **semantics** (the system of 'meaning') and **lexicogrammar** (the system of 'wording', including both grammar and vocabulary), and two expression strata – **phonology** and **phonetics** in the spoken mode, or **graphology** and **graphetics** in the written mode, or the equivalent two expression strata in a deaf sign language. This hierarchy of strata initially dominates the process of learning how to mean when we begin to learn a new language.

Learners have to learn language **trinocularly**, from all stratal angles (cf. Halliday 1996; Matthiessen and Halliday 1997): not only 'from below' (the resources of sounding and writing, or of signing, and then the resources of wording) and 'from above' (the resources of meaning in context), but also 'from within' (the internal organization of a given stratum as a linguistic subsystem). However, in language teaching, there has been a tendency to focus on one angle of approach to the (at least partial) exclusion of the others: the traditional grammar-based approach was 'from below'; the more recent notional syllabus approach provided a view 'from above', from the point of view of meaning at the semantic stratum; and the even more recent communicative and task-based approaches have also provided a view 'from above', but now from the point of view of meaning in context.

In current communicative approaches, one challenge faced by language teachers is thus how to relate the communicative approach 'from above' to the lower strata of language – to lexicogrammar in the first instance. Scholars involved in language teaching at universities are addressing this problem by developing systemic functional descriptions of the lexicogrammars of different languages (see Teruya, in press, on Japanese; Caffarel 2006; compare also the descriptions of different languages in Caffarel *et al.* 2004, and Li, forthcoming, on Chinese).

To meet the challenge of relating language to context, we need rich accounts of both lexicogrammar and semantics. On the one hand, we need semantics as an 'interface' between context and lexicogrammar (cf. Halliday 1973; Hasan 1984/1996) – a fully fledged semantics of text (see Martin 1992; Martin and Rose 2003). Language learners need to learn semantics as a ***strategic*** resource (cf. Halliday 1973) – a resource for transforming what is not meaning into meaning, construing their experience of the world as meaning and enacting social roles and relations as meaning; and this will provide them with the 'bridge' to lexicogrammar. On the other hand, both semantics and lexicogrammar need to be learned as ***resources*** rather than as ***inventories***. Thus, while the notional syllabus was important in focusing attention on learning how to mean, it presented notions as an inventory, failing to bring out the nature of semantics as a system, as pointed out by Gibbons and Markwick-Smith (1992). The local, internal organization of both semantics and lexicogrammar is thus critical in making them accessible to learners as resources: see Section 3.5.

To meet the challenge of relating language to context, we also need to have a systematic and comprehensive account of context (cf. Butt and Wegener, in press; Ghadessy 1999; Halliday 1978, 1992; Hasan 1980, 1985; Martin 1992, 1997). Like language, context is a semiotic system (see Halliday 1978), but it is a different kind of semiotic system, as shown by Martin (1992): language is a **denotative** semiotic system (that is, a semiotic system that has its own expression plane), whereas context is a **connotative** semiotic system (that is, a semiotic system that has other semiotic systems as its expression plane). Context is realized by language, and by other denotative semiotic systems such as gesture and facial expressions; and also by non-semiotic, social systems.

In learning a language, one does not learn features of stratal subsystems in isolation from one another; rather, one learns them as clusters of features that work together. This has been recognized in the language education literature. For example, Ur (1996) emphasizes the value of 'coordinating different language categories in a teaching programme', and she provides a table (p. 101) listing correspondences between 'situations', 'topics', 'notions and functions', 'grammar' and 'vocabulary'. These categories are related either stratally or metafunctionally, with a strong ideational orientation (thus, the situations are all characterized in field-like terms, and the 'topics' in ideational terms). This metafunctional imbalance can be brought out and adjusted in a systemic functional approach. In addition, the correspondences would be elaborated as strategies across strata rather than as lists of categories (cf. Matthiessen *et al.* 2005), and they can be fine-tuned as clusters for particular registers in particular situation types/institutional domains.

3.3 The spectrum of metafunction: learning different modes of meaning

3.3.1 Metafunctions as modes of meaning

The process of learning how to mean in a new language is multifunctional from the start since the process of making meaning is multifunctional in all languages. It involves the three metafunctional modes of meaning – **construing** our experience of the world around us and inside us as meaning (the **ideational** metafunction), **enacting** our social roles and relations as meaning (the **interpersonal** metafunction) and **presenting** the meanings construed and enacted as a flow of information (the **textual** metafunction) (see e.g., Halliday 1976, 1978, 1979; Halliday and Hasan 1985; Martin 1992, 1996; Matthiessen 2004, in press b).

These metafunctional modes of meaning are manifested in systems within the content plane of language – that is, within the semantic and lexicogrammatical strata; and they are reflected within the highest rank of the phonological stratum (that of the tone group). Table 1.2 provides an index into the most important systems in English, organized in terms of stratification and rank (for languages other than English, see e.g., Halliday and McDonald 2004: Table 6.2; Rose 2001; and Teruya, in press: Table 2.4; Caffarel 2006: Table 1.4; Matthiessen 2004: Table 10.1). The metafunctional organization of the structure of a clause is illustrated in Figure 1.2 below.

The balance between the interpersonal and the ideational metafunctions varies from one register to another (cf. Halliday 2001). For example, in terms of the text typology to be discussed below (see Figure 1.4 below), we can note that while expounding texts tend to be more ideationally oriented in overall organization, recommending texts tend to be more interpersonally oriented. However, this metafunctional balance also appears to vary somewhat across languages; this is one aspect of different 'fashions of speaking' or semantic styles. Such differences in 'metafunctional style' are quite subtle, and are thus a challenge even for advanced learners.

3.3.2 Resonance with context

The spectrum of functional diversity is also manifested within context as the three contextual variables of **field, tenor** and **mode**. Matthiessen *et al.* (2005) illustrates these for the situation characteristic of a certain type of telephonic service encounter.

The variables **resonate** with the three metafunctions in language (and their analogues in other semiotic systems operating within context alongside language): field resonates with ideational systems, tenor with interpersonal ones and mode with textual ones. The field, tenor and mode settings of a given situation type determine what ideational, interpersonal and textual meanings are 'at risk' in the register associated with that situation type (cf. Halliday 1978; Matthiessen 1993a). For example, in enabling contexts where people's behaviour is regulated (as opposed to enabling contexts where it is empowered), meanings that are to do with the modality of obligation are at risk, as are related meanings that are to do with responsibilities and privileges. For this reason, it makes sense to approach functional text typology 'from above' – from the vantage point of context: text types operate in situation types, and situation types are characterized as ranges of field, tenor and mode values. The text typology presented here (see Figure 1.4 below) is based on field and mode in the first instance, but tenor is equally important – it is just difficult to work out and diagram all combinations of field, tenor and mode values.

3.4 The cline of instantiation

3.4.1 The cline between system and text

The cline of instantiation extends from text **instances** to the overall systemic **potential** of a language – from the **acts of meaning** that make up a text to the

Table 1.2 Combined function–stratification matrix and function–rank matrix (lexicogrammar)

context			field			tenor	mode
	2nd-order		SEMIOTIC PROCESS, DOMAIN OF EXPERIENCE			SPEECH ROLE, VALUATION	MEDIUM, CHANNEL, RHETORICAL MODE — division of labour
	1st-order		SOCIAL PROCESS			INSTITUTIONAL ROLE, STATUS ROLE (POWER), SOCIOMETRIC ROLE (AFFECT), FAMILIARITY	
language			**ideational**			**interpersonal**	**textual**
			logical		**experiential**		
semantics	global		IDEATION			NEGOTIATION, HONORIFICATION/ POLITENESS	PERSPECTIVE
	paragraph		episodic patterns			exchange patterns	
	local		*[sequence:]* PROJECTION & EXPANSION		*[figure:]* CONFIGURATION	*[move:]* SPEECH FUNCTION, ASSESSMENT	*[message:]* SITING, FOCUSING
lexico-grammar	clause		TAXIS, LOGICO-SEMANTIC TYPE		TRANSITIVITY	MOOD, POLARITY, MODAL ASSESSMENT (incl. MODALITY)	THEME, INFORMATION [info. unit], VOICE; CONJUNCTION
	group	verbal		TENSE	EVENT TYPE, ASPECT	POLARITY, MODALITY	FOCUS
		nominal		MODIFICATION	THING TYPE, CLASSIFICATION, EPITHESIS, QUALIFICATION	PERSON, NOMINAL MOOD, ATTITUDE	REFERENCE
		adverbial		MODIFICATION	CIRCUMSTANCE TYPE	ATTITUDE TYPE	CONJUNCTION TYPE
phonology	tone group					TONE	TONICITY
			complex	unit			

rank	metafunction	axis				
		system	structure			
			it	*'s*	*lovely*	*darling*
clause	textual	unmarked theme, non-predicated, non-conjuncted	Theme	Rheme		
	interpersonal	free: declarative: non-tagged, vocative, non-interactant, temporal	Subject	Finite	Complement	Vocative
			Mood		Residue	
	experiential	relational: attributive & intensive	Carrier	Process	Attribute	
group			nominal group	verbal group	nominal group	nominal group
	interpersonal		non-interactant		non-interactant: positive reaction	non-interactant: positive affection

Figure 1.2 Metafunctional organization – lexicogrammar, ranks of clause and group

meaning potential that makes up the linguistic system. These are the outer poles of the cline; but between these poles there are intermediate patterns – patterns that we can interpret as subpotentials ('registers', 'genres') or as instance types ('text types').

The cline of instantiation is centrally involved in the learning of a language; learning means moving up and down the cline: learners 'distil' their own personal meaning potentials out of acts of meaning in text by moving up the cline, and they test this changing meaning potential in the instantiation of new acts of meaning, confirming or revising it (see Halliday 1992: 6–7, 1993). Here it is very helpful to think of language as 'languaging' in order to emphasize that it is both system and process and to avoid reifying it (see Halliday 1973; Swain, this volume).

As noted above, in the early stages of learning a new language, the hierarchy of stratification is the major challenge: learners must engage with all levels of stratification. However, as second/foreign language learners become more advanced, the process of learning how to mean can shift in focus from the dimension of stratification to the dimension of instantiation, as shown in Figure 1.3.

These two fundamental dimensions of organization are of course both part of the picture all along, but as learners become more advanced, the resources of the lower strata come into place as 'automated' realizations of semantics, allowing them to focus on meaning itself. More advanced learners can move further up the cline of instantiation towards the potential pole, learning more of the overall meaning potential of the language. (None of us ever make it all the way, even in our mother tongue, of course: the overall meaning potential is a **collective** resource, and we only operate with **personalized** sub-potentials of this collective meaning potential.)

3.4.2 Learning new registers in new situation types

This expansion of the advanced learner's meaning potential involves learning new registers (genres, text types) of the language – new functional varieties in new institutional settings – and the focus is thus increasingly on learning 'content' through language in these registers; compare Bhatia's (2004: 142–52) exploration of the relationship between 'generic competence' and 'professional expertise'. This learning is likely to involve **extensive** reading or indeed listening (rather than simply **intensive** reading or listening), with the advanced learners learning to process a greater quantity of text from a greater range of registers. It is likely to take place as learners take on new roles, both professional and non-professional, in different institutions outside the contexts of formal language education.

In a workplace, the registers that have to be learned include not only those associated with professional roles, but also those that are central to mateship in the creation of 'social capital' within the domain of tenor – like gossiping and teasing (cf. Eggins and Slade 1997), which can be quite hard for a person to learn in a foreign language operating in a foreign culture. Universities

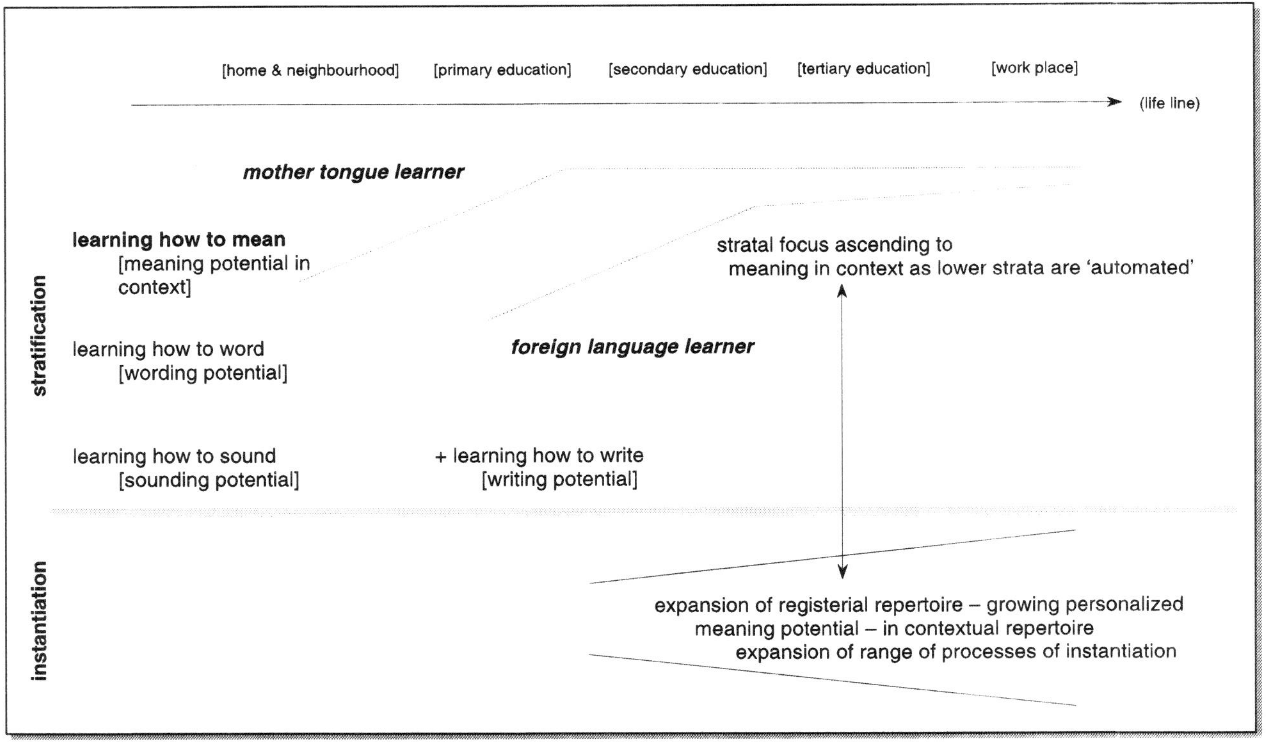

Figure 1.3 Stratification and instantiation in relation to learner's lifeline (ontogenesis)

represent a special case, of course: advanced learners often obtain degrees in a foreign language in a foreign country, so they have to learn the registers that go with such degrees. Training to be a translator or interpreter is a particularly interesting case, involving both a professional language-based skill and advanced language learning (for a systemic–functional investigation, see Kim, in preparation).

3.4.3 Context-based text/register typology/topology

Registers are functional varieties of language that can be characterized in terms of ranges of field, tenor and mode values in the contexts (situation types) within which registers are located (see Table 1.2 above for a list of the field, tenor and mode variables whose values define contexts): the diagram in Figure 1.4, based on a typological matrix designed by Ure, was devised by Teruya and me and focuses on field and mode. It is intended to suggest a **topological** interpretation of the context-based **typology** (see Martin and Matthiessen 1991 for the complementarity of typology and topology, and see Martin 2003 for a topology of registers occurring in history textbooks; compare also Bhatia's 2004 notion of genre colonies). The typology/topology in Figure 1.4 provides a **registerial map** where the texts advanced learners are presented with and have to engage with can be located.

The 'slices' of the display in Figure 1.4 represent the different socio-semiotic processes within field: **expounding** (general knowledge – explaining or documenting it), **reporting** (the flow of particular events), **recreating** (aspects of life, typically in an imaginary world), **sharing** (typically personal and particular experiences and personal values), **doing** (carrying out an activity sequence), **recommending** (a course of action), **enabling** (a course of action – empowering somebody to undertake it, or regulating their behaviour) and **exploring** (typically public issues and values).

The concentric circles in Figure 1.4 represent different mode values – from inner to outer circles: spoken and monologic, spoken and dialogic, written and dialogic, and written and monologic. Both the field and mode values shade into one another – as indicated by the topological display in Figure 1.4. For example, plays and screenplays constitute a mixed mode category in the sense that they are 'written to be spoken' (cf. Gregory 1967). New registers often emerge as mixed categories – like 'infomercials'; and registers typically evolve from other registers, crossing categorial boundaries in the typology/topology; for example, the academic research article (monologic) evolved out of letters between scholars (dialogic), as shown by Swales (1990), and modern news reports evolved out of traditional news stories (see e.g., Nanri 1993; Iedema, Feez and White 1994).

The representation of the context-based typology/topology in Figure 1.4 does not show all field and mode parameters. In terms of field, the socio-semiotic process is shown, but not the domain of experience – the 'subject matter' or 'topic area'. Texts clearly vary according to domain of experience; for example, texts in physics reflect the challenge of construing very complex sequences of events in physical systems as they are theorized today (see Unsworth 1995). However, the socio-semiotic process is a more important source of generalization than is the domain of experience; for example, explaining and documenting are strategies of expounding that generalize across different domains of experience, which is brought out when manifestations in different semiotic systems are considered. All these other field, tenor and mode dimensions must be imagined relative to Figure 1.4

In terms of mode, Figure 1.4 does not represent the different aspects that fall under the heading of 'rhetorical mode'. However, there are certain correlations: in 'doing' contexts, the role of the text in its context of situation is ancillary – it facilitates the performance of a non-semiotic, social process; in other contexts, the role of the text in its context is constitutive – it constitutes the performance of a semiotic process.

Tenor is not represented at all in Figure 1.4. We would need a sphere rather than a circle to represent tenor as well. However, the different regions of the circle tend to have different associations with tenor, and these can be mapped systematically. For example, in terms of VALUATION, sharing contexts are typically concerned with exploring similarities and differences in valuations among particular persons (often close friends – cf. Eggins 1990 – family and work mates); exploring contexts of the reviewing type are concerned with valuations of commercial goods-and-services in the public arena; and enabling contexts of the regulatory type are concerned with valuations of courses of action in terms of obligations. This means that different aspects of the resources of interpersonal ASSESSMENT are at risk.

The examples of registers given in Figure 1.4 are only illustrative, not exhaustive. They must be interpreted in terms of the location they are given in the typology/topology, not only in terms of their everyday sense. For example, prayers actually come in a number of different types (just like letters, as illustrated in the displays); prayers in recommending contexts are invocations, but there are also other types: prayers of celebration (doing), prayers of adoration (sharing) and prayers of meditation (expounding).

Most of the examples are of registers instantiated by fairly short homogeneous texts such as news reports and procedures. But a number of the examples are of **macro-registers** instantiated by long composite and heterogeneous texts such as textbooks, reference books, courses of lessons (cf. Christie 1997) and extended conversations (of the kind analysed in Eggins 1990). Such composite texts can be interpreted as rhetorical complexes of shorter texts (cf. Martin 1994) belonging to different registerial regions of the typology/topology. Texts falling within one region are typically 'nuclear' within such text complexes. For example, the nuclear texts within a textbook are 'expounding' in nature, but they are likely to be supported by 'exploring' texts (e.g., expositions of different interpretations), 'enabling' texts (e.g., procedures for lab experiments) and 'reporting' texts (e.g., historical background, including biographies of key scientists). Similarly, the nuclear texts within a newspaper are 'reporting' in nature, but newspapers consist of texts of a range of other types as well: see Iedema, Feez and White (1994).

Each region in the topological display in Figure 1.4 is the point of origin of a more delicate taxonomy. For example, one common form of 'expounding' in written monologic text is 'explaining', and the different strategies of explaining described by Veel (1997) – sequential explanation, causal explanation, factorial explanation, theoretical explanation and consequential explanation – can thus be located within the cell in the matrix defined by 'expounding', 'written' and 'monologic'.

Advanced learners may already have learned strategies located within the different regions of Figure 1.4, such as the strategies of explaining in their mother tongue, or they may be learning them at about the same time, in

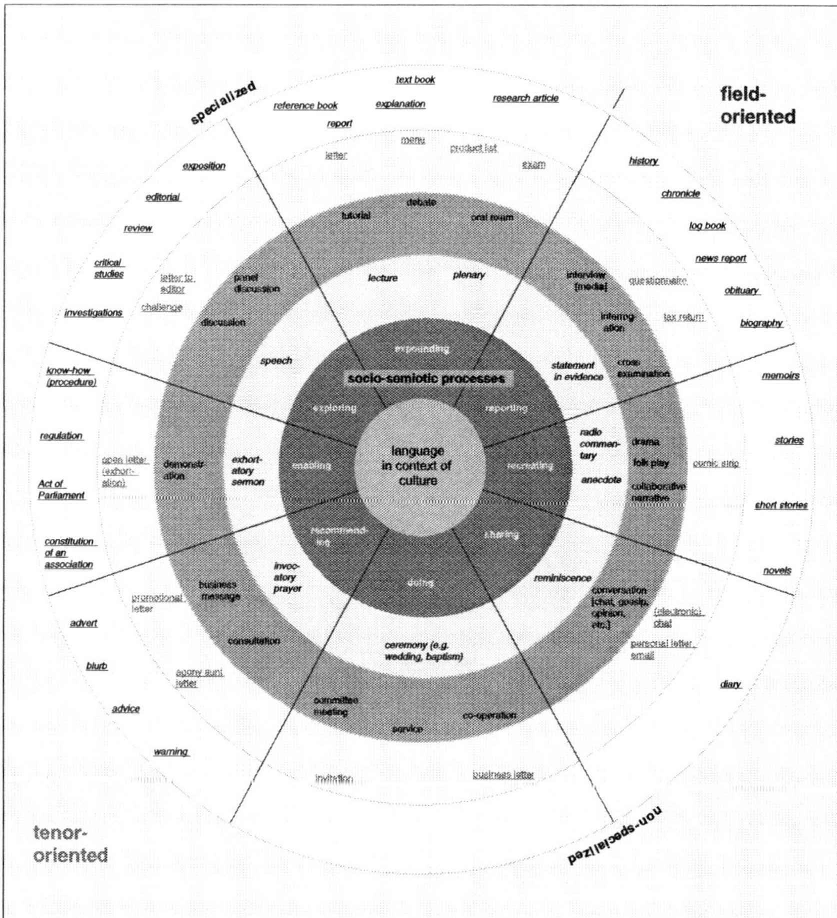

Figure 1.4 Context-based text typology/topology based on Jean Ure's taxonomy of texts

which case they can 'transfer' them to the foreign language. But it is also possible that the mother tongue and the foreign language they are learning differ in how they organize similar texts, as has been investigated in 'contrastive rhetoric' and in Rhetorical Structure Theory (see e.g., Abelen, Redeker and Thompson 1993, on Dutch and American English fund-raising letters; Trujillo Sáez 2001, on argumentative texts in Spanish and English). In either case, they are also learning grammatical features characteristic of the register of factorial explanations, including patterns of thematic progression, the use of explicit conjunctions and **grammatical metaphor** of the ideational type (see e.g., Halliday and Martin 1993; Halliday and Matthiessen 1999: Chapter 6, 2004: Chapter 10). Ideational grammatical metaphor will be central to many of the registers advanced learners engage with, as it is in the registers of science.

3.4.4 The cline of instantiation and semogenesis

The cline of instantiation makes it possible to differentiate and locate different kinds of **semogenesis** (processes of creating meaning), as shown in Figure 1.5 (cf. Halliday and Matthiessen 1999: 18). At the instance pole of the cline, semogenesis takes the form of **logogenesis**, the unfolding of the act of meaning as text. At the potential pole of the cline, it takes the form of **phylogenesis**, the evolution of the meaning potential in the human species. In between these two poles, semogenesis can be interpreted either from the vantage point of logogenesis as macro-logogenesis (a text type seen as a macro-text) or from the vantage point of phylogenesis as micro-phylogenesis (a register seen as a subpotential, as in Halliday's 1988, study of the evolution of scientific English since Chaucer, and in Nanri's 1993 study of the evolution of news reporting).

As noted above, **ontogenesis** (the development of a personalized meaning potential) involves accessing the meaning potential from the instance pole of the cline of instantiation through text. In this sense, it is a move up the cline of instantiation from the instance pole towards the collective potential pole. But individual meaners never reach this collective potential, and we can locate ontogenesis somewhere between the instance pole and the mid region of the cline: individual meaners will be characterized by their particular personal repertoire of registers – some selection out of the total pool constituting a particular language.

3.5 Local dimensions

The hierarchy of stratification, the spectrum of metafunction and the cline of instantiation together define the **global** organization of language in context. As we have seen, the cline of instantiation is the key to determining how far advanced learners can advance. While they can – and must – master all strata and all metafunctions, they cannot move all the way to the potential pole of the cline of instantiation. How far they advance along the cline of

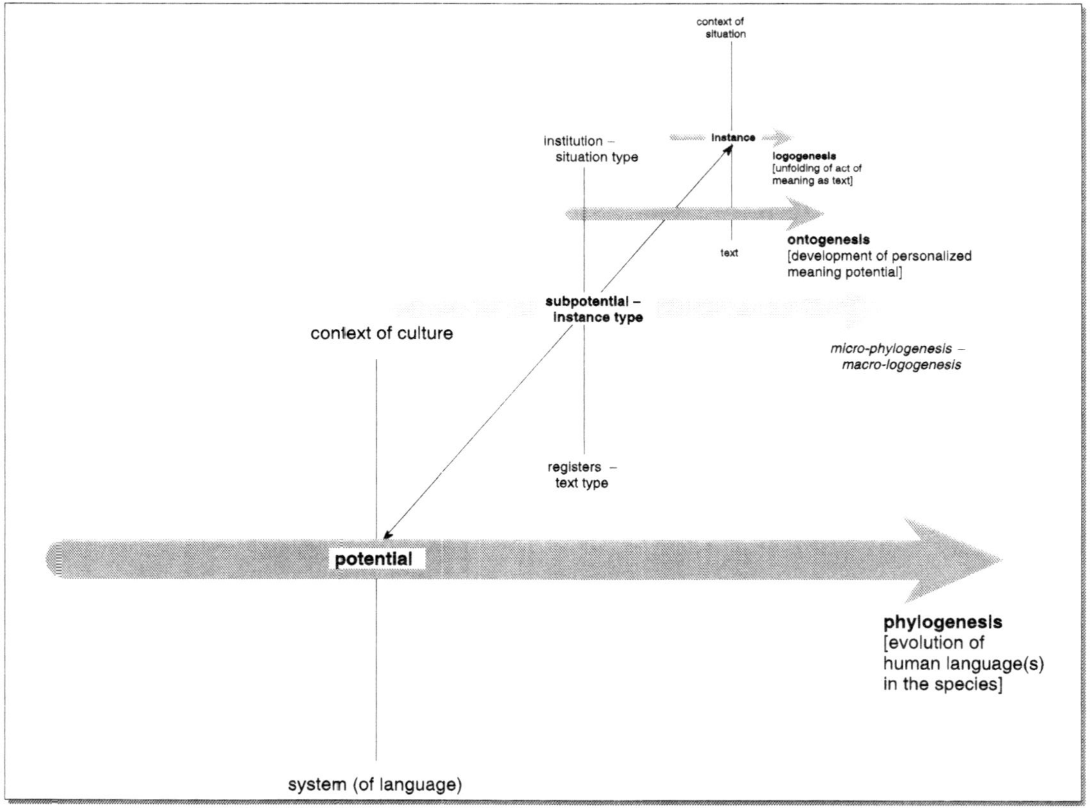

Figure 1.5 The three semogenic processes of phylogenesis, ontogenesis and logogenesis in relation to the cline of instantiation

instantiation will depend on the range of registers they come to master: a central goal for advanced language learners is to increase their registerial repertoires.

3.5.1 Register and systemic options 'at risk'

How significant the increase is can be seen in the local, systemic organization of lexicogrammar, semantics and context. (The dimensions **local** to a particular stratal subsystem – axis and rank – were introduced above; here I will focus on axis.) When context is represented systemically, we can see very clearly what the nature of the 'setting' of a particular situation type is and how it contrasts with other situation types in terms of the field, tenor and mode variables; when lexicogrammar and semantics are represented systemically as sets of interrelated options, we can identify resources of wording and meaning 'at risk' in the register associated with that situation type.[2] This will help us choose texts for advanced learners to enable them to expand their personal meaning potentials to include meanings at risk in a register.

In texts instantiating different registers, different options within the system of modality will be 'at risk'. For instance, in constitutions, within the general category of 'enabling' texts of the 'regulating' kind (cf. Figure 1.4), the combination of 'high' and 'obligation' is very common, and the modal auxiliary *shall*, which is uncommon in current English in general, is the registerial norm:

Text 1: Extract from a constitution

> CONSTITUTION FOR EPPING HEIGHTS PUBLIC SCHOOL PARENTS' AND CITIZENS' ASSOCIATION [. . .]
>
> 3. SCHOOL STAFF
>
> The Association **shall** not exercise any authority over the teaching staff or any matter relating to the control or management of the school. School staff **may** become members of the Association. The Principal of the school, or the Principal's nominee **shall** be a member, ex-officio, of the Association and all its committees.
>
> 4. MEMBERSHIP
>
> Membership **will** be open to all parents of pupils attending the school and to all citizens within the school community. The Association **shall** maintain a register of members. A person whose name appears in the register and who has paid the annual subscription **shall** be a member of the Association. The register **shall** be updated after each general meeting by the Secretary or the Secretary's nominee. If the name of a person has been omitted from the register when that person ***is*** otherwise ***entitled*** to be a member and their name **should** have been recorded in the register, then that person **shall** be a member of the Association.

In yet other registers, other options in modality are 'at risk'. For instance, in casual conversation within the general category of spoken, dialogic 'sharing' texts, explicitly subjective modalities of 'probability' are quite common; for example (from a dinner table conversation, the UTS/ Macquarie Corpus of spoken Australian English):

> But **I don't know** that we were friends. – Oh **I think** you were friends, you were friendly enough.
> **I think** it was ***probably*** right.
> **I think** there was ***probably*** a lot of truth in Prisoner.
> If you had a client like him that didn't have money you wouldn't be acting for him because **I don't think** there's any – **I don't think** there's any future in it.

The challenge for advanced learners is to master the full range of options operating in a given register. Gibbons and Markwick-Smith (1992) show how the systemic representation of MODALITY makes it possible to identify options in the system not selected by learners of English when they write essays even though native writers choose them regularly.

3.5.2 Systemic cartography

Representing a stratal subsystem, or part of such a subsystem such as the system of MODALITY in Figure 1.6, in the form of a system network thus has the advantage that it is possible to identify very clearly which options are 'at risk' in a given register in relation to the total potential represented by the network.[3] In general, by representing a stratal subsystem, we can map out the meaning potential or wording potential of that subsystem. This provides language educators with a much more useful account of the subsystem than a simple inventory of notions. We can also use the system network to locate a particular learner's or group of learners' **systemic frontier** relative to the overall systemic potential of the language. For example, a given learner may have mastered the modality systems of MODALITY TYPE and VALUE, but may be working with a restricted range within ORIENTATION and MANIFESTATION (see again Gibbons and Markwick-Smith 1992 for actual examples from their study).

Once we conceive of language as a resource – as a resource for making meaning – rather than as a rule system (cf. Halliday 1977), it follows that systemic organization is ***primary***, providing the ***environment*** for structural patterns (Halliday 1966, 1996). This is important because by using systemic organization, we can map out language holistically as a complex adaptive system (see e.g., Gell-Mann 1994), using a systems-thinking approach rather than a 'Cartesian Analysis' approach (cf. Capra 1996). The primacy of systemic organization is thus the key to the notion of mapping the resources of a language in comprehensive terms – which is of course what we need when we develop language education programs and curricula.

However, there are a number of other implications arising from treating

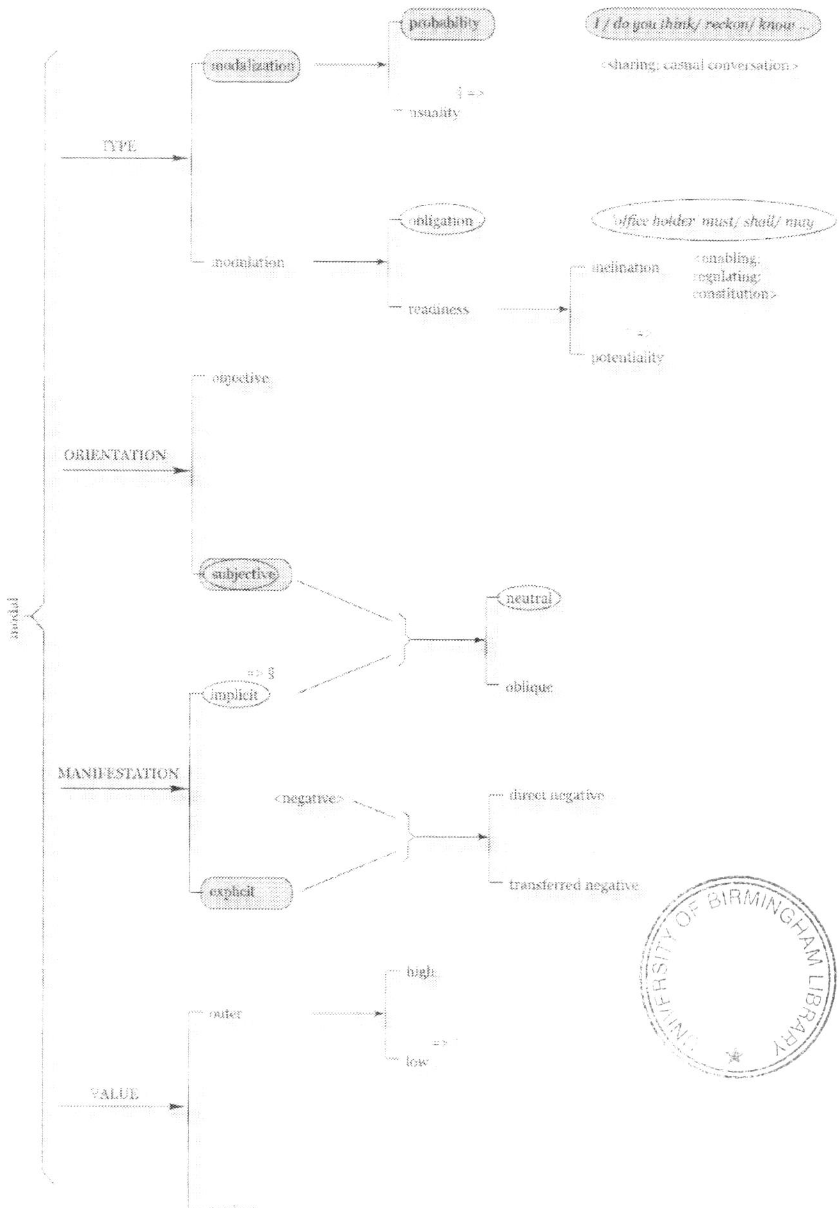

Figure 1.6 The interpersonal system of MODALITY, with indications of favoured selections in texts from two different registers (Shaded features pertain to casual conversation; circled features are common in the constitution text).

systemic organization as primary (cf. Halliday 1996; Matthiessen, in press a). Let me just mention two of these since they are quite central to the task of educating for advanced language capacities – (i) the grammar of intonation (see e.g., Halliday 1967; Halliday and Greaves, in press) and (ii) the lexicogrammatical continuum from grammar to lexis (see e.g., Hasan 1987; Matthiessen 1991; Tucker 1997).

4 Conclusion

In this chapter, I have discussed the theme of educating for advanced foreign language capacities in terms of systemic functional linguistics, noting the powerful momentum created by the resonance across a number of current approaches to language that are meaning-oriented, treat system and text as poles on a continuum, give language a central place among human systems, embody a dialogic perspective, and treat processes of meaning in context as data for linguistic description, theorizing and application. To develop advanced foreign language education further, we must understand the central phenomenon – language in context. Any theory of language learning and any applications in educational processes must be informed by a deep insight into the key properties of language. Indeed, as Halliday (1993) has suggested, developing a language-based theory of learning in general is both possible and desirable.

Following Halliday and other systemic functional linguists, I have explored language as a meaning-making system, interpreting it as a resource organized in terms of a number of semiotic dimensions and characterizing language learning as a multidimensional process. Dealing with these dimensions one at a time, I have discussed what they reveal about the nature of language and what the implications are for an exploration of advanced foreign language education.

Notes

1 'Making meaning' includes the process of communicating, in the exchange of meaning between speaker and addressee, but it is broader than 'communication'. The notion of 'making meaning' is a constructivist one (cf. Halliday and Matthiessen 1999), relating to ideas by Whorf, Vygotsky and Bakhtin. As noted in Section 1, this is very different from the common position Whorf called 'natural logic' – a position that is often linked to the conception of language as a tool for communicating pre-linguistic ideas.

2 One way of showing what meanings are 'at risk' in a given register is to set up a register-specific system (see Halliday 1973): Caffarel (1992) shows how the grammatical system of tense in French is deployed semantically by a number of registerially distinct semantic tense systems.

3 As Halliday has always emphasized, this is of course a matter of degree – the probability that one term or another in a system will be instantiated in a text within a given register; for illustrations based on counts in texts, see e.g., Matthiessen (2002, 2006). One can thus ask to what extent a given learner

has the same probability profile in relation to a system such as MODALITY in a given register as native speakers do.

References

Abelen, E., Redeker, G. and Thompson, S. A. (1993) 'The rhetorical structure of US-American and Dutch fund-raising letters'. *Text*, 13, 323–50.

Bhatia, V. K. (2004) *Worlds of Written Discourse*. London: Continuum.

Butt, D. G. and Wegener, R. (in press) 'The work of concepts: "context" and "metafunction" in the systemic functional model', in R. Hasan, C. M. I. M. Matthiessen and J. Webster (eds), Volume 2.

Caffarel, A. (1992) 'Interacting between a generalized tense semantics and register-specific semantic tense systems: a bi-stratal exploration of the semantics of French tense'. *Language Sciences*, 14, 385–418.

Caffarel, A. (2006) *A Systemic Functional Grammar of French: From Grammar to Discourse*. London: Continuum.

Caffarel, A., Martin, J. R. and Matthiessen, C. M. I. M. (eds) (2004) *Language Typology: A Functional Perspective*. Amsterdam/Philadelphia: John Benjamins.

Capra, F. (1996) *The Web of Life: A New Synthesis of Mind and Matter*. London: Harper Collins.

Christie, F. (1997) 'Curriculum macrogenres as forms of initiation into a culture', in F. Christie and J. R. Martin (eds), pp. 134–60.

Christie, F. and Martin, J. R. (eds) (1997) *Genre and Institutions: Social Processes in the Workplace and School*. London: Cassell.

Eggins, S. (1990) 'Conversational Structure: A Systemic-functional Analysis of Interpersonal and Logical Meaning in Multiparty Sustained Talk'. University of Sydney: Unpublished Ph.D. dissertation.

Eggins, S. and Slade, D. (1997) *Analysing Casual Conversation*. London: Cassell.

Gell-Mann, M. (1994) *The Quark and the Jaguar: Adventures in the Simple and the Complex*. London: Abacus.

Ghadessy, M. (ed) (1999) *Text and Context in Functional Linguistics*. Amsterdam/Philadelphia: Benjamins.

Gibbons, J. and Markwick-Smith, V. (1992) 'Exploring the use of a systemic semantic description'. *International Journal of Applied Linguistics*, 2, 36–51.

Gregory, M. J. (1967) 'Aspects of varieties differentiation'. *Journal of Linguistics*, 3, 177–98.

Halliday, M. A. K. (1966) 'Some notes on "deep" grammar'. *Journal of Linguistics*, 2, 57–67.

Halliday, M. A. K. (1967) *Intonation and Grammar in British English*. The Hague: Mouton (Janua Linguarum Series Practica 48).

Halliday, M. A. K. (1973) *Explorations in the Functions of Language*. London: Edward Arnold.

Halliday, M. A. K. (1975) *Learning How to Mean*. London: Edward Arnold.

Halliday, M. A. K. (1976) *System and Function in Language*. London: Oxford University Press.

Halliday, M. A. K. (1977) 'Ideas about language', in M. A. K. Halliday, *Aims and Perspectives in Linguistics*. Applied Linguistics Association of Australia (Occasional Papers 1), 32–49.

Halliday, M. A. K. (1978) *Language as Social Semiotic: The Social Interpretation of Language and Meaning*. London and Baltimore: Edward Arnold and University Park Press.

Halliday, M. A. K. (1979) 'Modes of meaning and modes of expression: types of grammatical structure and their determination by different semantic functions', in D. J. Allerton, E. Carney and D. Holdcroft (eds), *Function and Context in Linguistic Analysis: A Festschrift for William Haas.* Cambridge: Cambridge University Press, pp. 57–79.

Halliday, M. A. K. (1980) 'Three aspects of children's language development: learning language, learning through language, learning about language', in M. M. Haussler, D. S. Strickland and Y. M. Goodman (eds), *Oral and Written Language Development: Impact on Schools.* (International Reading Association & National Council of Teachers of English: Proceedings from the 1979/1980 IMPACT Conferences), pp. 7–19.

Halliday, M. A. K. (1988) 'On the language of physical science', in M. Ghadessy (ed.), *Registers of Written English: Situational Factors and Linguistic Features.* London & New York: Pinter Publishers, pp. 162–78.

Halliday, M. A. K. (1992) 'The notion of "context" in language education', in T. Le and M. McCausland (eds), *Interaction and Development: Proceedings of the International Conference,* Vietnam, 30 March–1 April 1992. University of Tasmania: Language Education, pp. 1–26.

Halliday, M. A. K. (1993) 'Towards a language-based theory of learning'. *Linguistics and Education,* 5, 93–116.

Halliday, M. A. K. (1995) 'On language in relation to the evolution of human consciousness', in S. Allén (ed.), *Of Thoughts and Words: Proceedings of Nobel Symposium 92* 'The relation between language and mind', Stockholm, 8–12 August 1994. Singapore, River Edge NJ and London: Imperial College Press, pp. 45–84.

Halliday, M. A. K. (1996) 'On grammar and grammatics', in R. Hasan, C. Cloran and D. Butt (eds), *Functional Descriptions: Theory into Practice.* Amsterdam/Philadelphia: John Benjamins, pp. 1–38.

Halliday, M. A. K. (2001) 'On the grammatical foundations of discourse', in R. Shaozeng, W. Guthrie and I. W. R. Fong (eds), *Grammar and Discourse: Proceedings of the International Conference on Discourse Analysis,* University of Macau (in conjunction with Tsinghua University, China), 16–18 October 1997. Macau: University of Macau Publication Centre, pp. 47–57.

Halliday, M. A. K. (2002) 'Computing meanings: some reflections on past experience and present prospects', in G. Huang and Z. Wang (eds), *Discourse and Language Functions.* Shanghai: Foreign Language Teaching and Research Press, pp. 3–25.

Halliday, M. A. K. (2003) *The Language of Early Childhood.* Volume 4: The Collected Works of M. A. K. Halliday, ed. by J. Webster. London: Continuum.

Halliday, M. A. K. and Greaves, W. S. (in press) *Intonation in the Grammar of English.* London: Equinox.

Halliday, M. A. K. and Hasan, R. (1985) *Language, Context and Text: A Social Semiotic Perspective.* Geelong, Australia: Deakin University Press.

Halliday, M. A. K. and Martin, J. R. (1993) *Writing Science: Literacy and Discursive Power.* London: Falmer.

Halliday, M. A. K. and Matthiessen, C. M. I. M. (1999) *Construing Experience through Meaning: A Language-based Approach to Cognition.* London: Cassell.

Halliday, M. A. K. and Matthiessen, C. M. I. M. (2004) *An Introduction to Functional Grammar* (3rd rev. edn). London: Hodder Arnold.

Halliday, M. A. K. and McDonald, E. (2004) 'Metafunctional profile of the grammar of Chinese', in A. Caffarel, J. R. Martin and C. M. I. M. Matthiessen (eds), pp. 305–96.

Hasan, R. (1980) 'What's going on: a dynamic view of context', in J. E. Copeland and

P. W. Davis (eds), *The Seventh LACUS Forum*. Columbia, SC: Hornbeam Press, pp. 106–21.
Hasan, R. (1984/1996) 'The nursery tale as a genre', reprinted in Hasan (1996), pp. 51–72.
Hasan, R. (1985) 'Meaning, context and text: fifty years after Malinowski', in J. D. Benson and W. S. Greaves (eds), *Systemic Perspectives on Discourse*. Norwood, NJ: Ablex, pp. 16–50.
Hasan, R. (1987) 'The grammarian's dream: lexis as most delicate grammar', in M. A. K. Halliday and R. P. Fawcett (eds), *New Developments in Systemic Linguistics: Theory and Description*. London: Pinter, pp. 184–211.
Hasan, R. (1996) *Ways of Saying: Ways of Meaning. Selected Papers of Ruqaiya Hasan*. C. Cloran, D. Butt and G. Williams, eds. London: Cassell.
Hasan, R. and Martin, J. R. (eds) (1989) *Language Development: Learning Language, Learning Culture. Meaning and Choice in Language: Studies for Michael Halliday*. Norwood, NJ: Ablex.
Hasan, R., Matthiessen, C. M. I. M. and Webster, J. (eds) (2005) *Continuing Discourse on Language: A Functional Perspective*. 2 volumes. London: Equinox.
Iedema, R., Feez, S. and White, P. (1994) *Media Literacy*. (Write it right industry research report no. 2.) Sydney: NSW, Department of Education, Disadvantaged Schools Program Metropolitan East.
Kim, M. (in preparation) 'Application of systematic-functional grammar-based text analysis in translator education: A case study of English/Korean translation'. Sydney: Macquarie University. Unpublished Ph.D. thesis.
Lado, R. (1957) *Linguistics across Cultures: Applied Linguistics for Language Teachers*. Ann Arbor, MI: The University of Michigan Press.
Lantolf, J. P. and Thorne, S. L. (2006) *Sociocultural Theory and the Genesis of Second Language Development*. Oxford: Oxford University Press.
Li, E. (forthcoming) *A Systemic Functional Grammar of Chinese*. London: Continuum.
Martin, J. R. (1988) 'Grammatical conspiracies in Tagalog: family, face and fate – with reference to Benjamin Lee Whorf', in M. J. Cummings, W. S. Greaves and J. D. Benson (eds), *Linguistics in a Systemic Perspective*. Amsterdam: John Benjamins, pp. 243–300.
Martin, J. R. (1992) *English Text: System and Structure*. Amsterdam: John Benjamins.
Martin, J. R. (1994) 'Macro-genres: the ecology of the page'. *Network 21*. Newsletter with news, views and reviews in systemic linguistics and related areas, pp. 29–52, available on line at *http://minerva.ling.mq.edu.au/Resources/Network/Network.html*.
Martin, J. R. (1996) 'Types of structure: deconstructing notions of constituency in clause and text', in E. Hovy and D. Scott (eds), *Burning Issues in Discourse: A Multidisciplinary Perspective*. Heidelberg: Springer, pp. 39–66.
Martin, J. R. (1997) 'Analysing genre: functional parameters', in F. Christie and J. R. Martin (eds), pp. 3–39.
Martin, J. R. (2003) 'Making history: grammar for interpretation', in J. R. Martin and R. Wodak (eds), *Re/reading the Past: Critical and Functional Perspectives on Time and Value*. Amsterdam/Philadelphia: John Benjamins, pp. 19–57.
Martin, J. R. and Matthiessen, C. M. I. M. (1991) 'Systemic typology and topology', in F. Christie (ed.), *Literacy in Social Processes: Papers from the Inaugural Australian Systemic Functional Linguistics Conference*, Deakin University, January 1990. Darwin: Centre for Studies of Language in Education, Northern Territory University, pp. 345–83.
Martin, J. R. and Rose, D. (2003) *Working with Discourse*. London: Continuum.

Matthiessen, C. M. I. M. (1991) 'Lexico(grammatical) choice in text-generation', in C. Paris, W. Swartout and W. C. Mann (eds), *Natural Language Generation in Artificial Intelligence and Computational Linguistics*. Boston: Kluwer, pp. 249–92.

Matthiessen, C. M. I. M. (1993a) 'Register in the round: diversity in a unified theory of register analysis', in M. Ghadessy (ed.), London: Pinter, pp. 221–92.

Matthiessen, C. M. I. M. (1993b) 'The object of study in cognitive science in relation to its construal and enactment in language', in *Language as Cultural Dynamic* (Special issue of Cultural Dynamics VI.1–2: pp. 187–243, ed. M. A. K. Halliday).

Matthiessen, C. M. I. M. (1995) *Lexicogrammatical Cartography: English Systems*. Tokyo: International Language Sciences Publishers.

Matthiessen, C. M. I. M. (1998) 'Construing processes of consciousness: from the commonsense model to the uncommonsense model of cognitive science', in J. R. Martin and R. Veel (eds), *Reading Science: Research, Popular Culture, Industry and Schooling*. London: Routledge, pp. 327–57.

Matthiessen, C. M. I. M. (2002) 'Combining clauses into clause complexes: a multi-faceted view', in J. Bybee and M. Noonan (eds), *Complex Sentences in Grammar and Discourse: Essays in Honor of Sandra A. Thompson*. Amsterdam/Philadelphia: John Benjamins, pp. 237–322.

Matthiessen, C. M. I. M. (2004) 'Descriptive motifs and generalizations', in A. Caffarel, J. R. Martin and C. M. I. M. Matthiessen (eds), pp. 537–673.

Matthiessen, C. M. I. M. (2006) 'Frequency profiles of some basic grammatical systems: an interim report', in S. Hunston and G. Thompson (eds), *System and Corpus: Exploring Connections*. London: Equinox, pp. 103–42.

Matthiessen, C. M. I. M. (in press a) 'The "architecture" of language according to systemic functional theory: developments since the 1970s', in R. Hasan, C. M. I. M. Matthiessen and J. Webster (eds).

Matthiessen, C. M. I. M. (in press b) 'Lexicogrammar in systemic functional linguistics: descriptive and theoretical developments in the "IFG" tradition since the 1970s', in R. Hasan, C. M. I. M. Matthiessen and J. Webster (eds).

Matthiessen, C. M. I. M., Butt, D. G., Cleirigh, C., Lukin, A. and Nesbitt, C. (2005) 'Welcome to Pizza Hut: a case study of multistratal analysis', *Australian Review of Applied Linguistics*, 19, 123–50.

Matthiessen, C. M. I. M. and Halliday, M. A. K. (1997) 'Systemic functional grammar: a first step into the theory'. Available online: http://minerva.ling.mq.edu.au/ resource/ VirtuallLibrary/Publications/sfg_firststep/SFG%20intro%20New.html.

Mohan, B. A. (1979) 'Relating language teaching and content teaching'. *TESOL Quarterly*, 13, 171–82.

Mohan, B. A. (1986) *Language and Content*. Reading, MA: Addison-Wesley.

Mohan, B. A. (1989) 'A situation-based approach to language socialization'. *Word*, 40, 99–117.

Mohan, B. A. and Zingzi, H. (2002) 'Assessing the integration of language and content in mandarin as a foreign language classroom'. *Linguistics and Education*, 13, 405–33.

Nanri, K. (1993) 'An attempt to synthesize two systemic contextual theories through the investigation of the process of the evolution of the discourse semantic structure of the newspaper reporting article'. University of Sydney: Unpublished Ph.D. thesis.

Ortega, L. and Byrnes, H. (eds) (2007) *The Longitudinal Study of Advanced L2 Capacities*. Mahwah, NJ: Lawrence Erlbaum.

Painter, C. (1999) *Learning through Language in Early Childhood*. London: Cassell.

Rose, D. (2001) *The Western Desert Code: An Australian Cryptogrammar.* Canberra: Pacific Linguistics.

Steiner, E. (1991) *A Functional Perspective on Language, Action and Interpretation.* Berlin/New York: Mouton de Gruyter.

Stutterheim, von C. and Carroll, M. (2006) 'The impact of grammatical temporal categories on ultimate attainment in L2 learning', in H. Byrnes, H. Weger-Guntharp and K. A. Sprang (eds), *Educating for Advanced Foreign Language Capacities: Constructs, Curriculum, Instruction, Assessment.* Washington, DC: Georgetown University Press, pp. 40–53.

Swales, J. (1990) *Genre Analysis: English in an Academic Research Setting.* Cambridge: Cambridge University Press.

Teruya, K. (in press) *Systemic Functional Grammar of Japanese.* 2 volumes. London: Continuum.

Trujillo Sáez, F. (2001) 'An analysis of argumentative texts for contrastive rhetoric', in D. Levey, M. A. Losey and M. A. González (eds), *English Language Teaching: Changing Perspectives in Context.* Cádiz: Servicio de Publicaciones de la Universidad de Cádiz, pp. 493–505.

Tucker, G. H. (1997) *The Lexicogrammar of Adjectives: A Systemic Functional Approach to Lexis.* London: Cassell.

Unsworth, L. (1995) 'How and Why: Recontextualizing Science Explanations in School Science Books'. Sydney University: Unpublished Ph.D. dissertation.

Ur, P. (1996) *A Course in Language Teaching: Practice and Theory.* Cambridge: Cambridge University Press.

Veel, R. (1997) 'Learning how to mean – scientifically speaking: apprenticeship into scientific discourse in the secondary school', in F. Christie and J. R. Martin (eds), pp. 161–95.

Wells, G. (1999) *Dialogic Inquiry: Toward a Sociocultural Practice and Theory of Education.* Cambridge: Cambridge University Press.

Whorf, B. L. (1956) *Language, Thought and Reality,* edited by J. B. Carroll. Cambridge, MA: MIT Press.

Williams, G. (2005) 'Grammatics in schools', in R. Hasan, C. M. I. M. Matthiessen and J. Webster (eds), pp. 281–310.

Wu, C. (2000) 'Modelling Linguistic Resources'. Sydney: Macquarie University. Unpublished Ph.D. thesis.

2 Generalized collective dialogue and advanced foreign language capacities[1]

James V. Wertsch

Discussions of advanced language capacity are often couched in terms of language and culture, a formulation that begs the question of how these two notions are related. Is language part of culture? Is culture part of language? Where does one end and the other begin? Anyone who has dealt with these questions knows that they have no simple answers. It is possible, however, to say something about the assumptions that theoretical traditions bring to the table when these issues are discussed, and this will be my topic in what follows.

My focus will be on a tradition of scholarship about language, thought and culture that provides a particular perspective on this issue. One of the defining aspects of this tradition is what might be called its 'linguistic imperialism', or the tendency to incorporate phenomena as part of language that other approaches would include under the heading of 'culture'. In today's parlance the approach I have in mind might be viewed as a version of sociolinguistics, pragmatics or discourse analysis, but it has somewhat different intellectual roots than those that guide many endeavours in these fields. Specifically, it comes from the writings of two twentieth-century Russian scholars: Mikhail Mikhailovich Bakhtin (1895–1975) and Gustav Gustavovich Shpet (1879–1937).

Bakhtin on text

My starting point is Bakhtin's account of text.[2] In an article 'The Problem of the Text in Linguistics, Philology and the Human Sciences: An Experiment in Philosophical Analysis', Bakhtin outlined 'two poles' of text.

> Each text presupposes a generally understood (that is, conventional within a given collective) system of signs, a language (if only the language of art) . . . And so behind each text stands a language system. Everything in the text that is repeated and reproduced, everything repeatable and reproducible, everything that can be given outside a given text (the given) conforms to this language system. But at the same time each text (as an utterance) is individual, unique, and unrepeatable, and

herein lies its entire significance (its plan, the purpose for which it was created) . . . With respect to this aspect, everything repeatable and reproducible proves to be material, a means to an end . . . The second aspect (pole) inheres in the text itself, but is revealed only in a particular situation and in a chain of texts (in the speech communication of a given area). (1986c: 105)

Bakhtin is best known for his theory of the utterance, a concern that is reflected in the assertion that the 'entire significance [of a text] (its plan, the purpose for which it was created)' can be traced to its 'individual, unique, and unrepeatable' pole. In what follows, however, I shall focus largely on the other pole of text, the one concerned with 'repeatable and reproducible' elements provided by a 'language system' that is 'conventional within a given collective'.

The first inclination of those of us coming from traditions of contemporary linguistics in the United States is to understand what Bakhtin referred to as a 'language system' in terms of standard treatments of morphology, syntax and semantics. This, however, would be more a reflection of our perspective than that of Bakhtin or his translators, and for this reason it should be resisted. Instead, Bakhtin had in mind an account of the repeatable and reproducible pole of text that recognizes these elements – but also includes a second level of organization in a 'language system' and a corresponding second level of analysis. In this view the first level for analysing a language system has to do with the structural analysis of decontextualized sentences and the second level focuses on 'social languages', 'speech genres' and the 'chain of texts' in which an utterance appears.

Presenting Bakhtin's ideas from a perspective more familiar to Western readers, Michael Holquist has formulated this point as follows:

'Communication' as Bakhtin uses the term does indeed cover many of the aspects of Saussure's *parole*, for it is concerned with what happens when real people in all the contingency of their myriad lives actually speak to each other. But Saussure conceived the individual language user to be an absolutely free agent with the ability to choose any words to implement a particular intention. Saussure concluded, not surprisingly, that language as used by heterogeneous millions of such willful subjects was unstudiable, a chaotic jungle beyond the capacity of science to domesticate. (1986: xvi)

Accepting a stark Saussurean opposition has clear implications for advanced language acquisition. Specifically, it suggests that the task is to get students to master a set of rules of *langue*, and then assume that the appropriate use of language forms involves some combination of individual choice and cultural context. In short, issues of language use and how utterances are shaped by their positioning in a 'chain of texts' fall outside the framework of what is properly considered language.

As Holquist (1986) points out, however, one of Bakhtin's insights was that the semiotic world need not be divided up so starkly as the *langue-parole* distinction suggests. In this regard Bakhtin wrote:

> the single utterance, with all its individuality and creativity, can in no way be regarded as a *completely free combination* of forms of language, as is supposed, for example by Saussure (and by many other linguists after him), who juxtaposed the utterance *(la parole),* as a purely individual act, to the system of language as a phenomenon that is purely social and mandatory for the individuum. (Bakhtin 1986b: 81)

Instead, as Holquist notes, 'Bakhtin . . . begins by assuming that individual speakers do not have the kind of freedom *parole* assumes they have. The problem here is that the great Genevan linguist overlooks the fact that "in addition to the forms of languages there are also *forms of combinations* of these forms"' (1986: xvi).

What Bakhtin has to say about these forms of combinations of forms amounts to a call for a second level of analysis *within* the pole of text having to do with what is 'repeated and reproduced'. It expands what needs to be taken into account when talking about a 'language system' or 'a generally understood (that is, conventional within a given collective) system of signs.'

The key to understanding the implications of Bakhtin's insights on these issues is his concept of 'dialogism' and the related notions of 'voice' and 'multivoicedness'. Throughout his writings Bakhtin emphasized that a defining property of utterances is that they exist only in dialogic contact with other utterances and hence are 'filled with *dialogic overtones*' (1986b: 102). It is this dialogic contact that provides the key to understanding the second level of phenomena within Bakhtin's second pole of text.

Key to understanding this issue is Bakhtin's assumption that the word 'never' belongs solely to the speaker; instead, is it always 'half someone else's' (1981: 293), the result being the inherent multivoicedness of utterances.

> [The word] becomes 'one's own' only when the speaker populates it with his own intention, his own accent, when he appropriates the word, adapting it to his own semantic and expressive intention. Prior to this moment of appropriation, the word does not exist in a neutral and impersonal language (it is not, after all, out of a dictionary that the speaker gets his words!), but rather it exists in other people's mouths, in other people's concrete contexts, serving other people's intentions: it is from there that one must take the word, and make it one's own. (Bakhtin 1981: 293–4)

When dealing with utterances from the perspective of Bakhtin's first pole of text, contemporary sociolinguistic analyses have little trouble making sense of the phenomena involved. For example, his claims are consistent with analyses of how utterances can be co-constructed or how they can be abbreviated responses to a question (Speaker 1: 'What time is it?' Speaker 2: 'Two o'clock.').

What is significant, however, is that Bakhtin saw the claim about words being half someone else's as applying to *language* – not text or utterance. And this brings us back to a level of analysis that goes beyond the categories of *langue* and *parole.* Specifically, it involves a level of language phenomena that exist as collectively shared social facts about the organization of utterances, on the one hand, but are not reducible to standard accounts of grammatical categories, on the other.

In an effort to get at what Bakhtin had in mind in this regard, I shall introduce a distinction between 'local dialogue' and 'generalized collective dialogue'. Local dialogue is what Bakhtin sometimes called the 'primordial dialogism of discourse' (1981: 275) and involves ways in which one speaker's concrete utterances come into contact with, or 'interanimate', the utterances of another. This form of dialogic interanimation involves 'direct, face-to-face vocalized verbal communication between persons' (Voloshinov 1973: 95) and is what usually comes to mind first when we encounter the term 'dialogue'.

For Bakhtin, however, the voices of multiple speakers come into contact at the level of generalized collective dialogue as well, and this leads to additional ways in which words can be 'filled with *dialogic overtones*' (1986b: 102). The notion of generalized collective dialogue concerns how utterances reflect the voice of others, including entire groups, who are not present in the immediate speech situation.

From his writings it is clear that Bakhtin recognized something like the distinction I am outlining. He viewed dialogue as ranging from the face-to-face primordial dialogue of discourse noted above, which falls under the heading of what I am calling localized dialogue, to ongoing, potentially society-wide interchanges, which fall under the heading of generalized collective dialogue. An addressee can be

> an immediate participant-interlocutor in an everyday dialogue, a differentiated collective of specialists in some particular area of cultural communication, a more or less differentiated public, ethnic group, contemporaries, like-minded people, opponents and enemies, a subordinate, a superior, someone who is lower, higher, familiar, foreign, and so forth. And it can also be an indefinite, unconcretized *other*. (Bakhtin 1986b: 95)

When discussing language acquisition, the implications of local dialogue are clear. What would it mean, after all, to say that someone knows a language but has not mastered the indexical and other pragmatic devices needed to link her utterances to those of others? Indeed, some would argue that such pragmatic issues are the true foundation of learning a language.

But when it comes to the kind of extended, society-wide dialogue Bakhtin mentions, many would question whether this is a matter of *language* acquisition. To be sure, being a competent member of a *culture or society* or some other nonlinguistic grouping might require such knowledge, but in what sense does it touch on what it means to speak a language correctly?

Shpet on the inner form of the word

In an attempt to answer this question, I turn to some ideas of Shpet. To begin, a few words about the individual. After studying with Husserl and bringing phenomenology to Russia in the 1910s, Shpet became one of the leading philosophers and hermeneutic phenomenologists in the USSR, but this did little to spare him a horrible fate; indeed, it may have hastened it. On several occasions he was dismissed from research and teaching positions, and in 1935 he was arrested and sent to Eniseisk in Siberia. In 1937 he was arrested again and a 'troika' of the Narodnyi Komissariat Vnutrennikh Del (People's Commissariat of Internal Affairs) sentenced him to ten years in prison without the right to correspond with anyone. Later that same year he was once again arrested, brutally interrogated, sentenced and executed. In 1956 the rehabilitation of Shpet began, but in certain respects it continues to this day.

Shpet produced several volumes of work, ranging from his *Yavlenie i Smysl'* (Appearance and Sense) in 1914, which is credited with introducing phenomenology to Russia to works on psychology, aesthetics, philosophy and literature. In what follows I shall focus on his fascinating – and in some sections, nearly impenetrable – 1927 volume *Vnutrennyaya Forma Slova*[3]*: Etyudy i Variatsii na Temy Gumbol'dta* (The Inner Form of the Word: Studies and Variations on a Humboldtian Theme).

Because of his troubles with the authorities, all traces of Shpet, not to mention of his writings, virtually disappeared during the Soviet period. However, *The Inner Form of the Word* had a powerful influence on many readers and students, including Lev Semënovich Vygotsky, who attended Shpet's seminars for two years. This is not to say that this influence was explicitly acknowledged. Vygotsky, for example, almost never cited Shpet, probably because he knew he could not do so in the dangerous political context in which he was living. The influence is none the less obvious. For example, some sections of Chapter 7 of Vygotsky's *Thinking and Speech* (1987, originally published in 1934) appear to be taken almost verbatim out of *The Inner Form of the Word.*

A starting point for Shpet is that the word must be understood as a living, organic entity, a view that contrasts markedly with understanding it as a decontextualized item that appears in a dictionary or in an abstract grammar. In this connection Shpet asserted that, 'We must look at language not as a dead product of a generative process (ein Erzeugtes), but instead as a generative process (eine Erzeugung)' (1996: 55). This is a central tenet of the argument he lays out in *The Inner Form of the Word,* where Shpet repeats at several points Humboldt's position that language must be viewed as *energeia* and not as *ergon.*

> [F]or Humboldt it was a major revelation that language is energeia. For him everything comes down to this. All the other nuances in the description of this term must be understood in this sense: language is 'activity of the spirit' and 'the immanent work of the soul'. It is at the foundation of the very nature of being human . . . Language is a social thing and a psychophysiological process, but it is also exists as an idea. Language can be viewed not only as substance, but as subject, not only as a thing, product, or result of production, but as production process, as energeia. (1996: 77–8)

The notion that language is a subject, an activity of the spirit, adds an element of dynamism that is often not a part of contemporary Western traditions of scholarship.

Humboldt's legacy usually assumes that language as subject is a notion that applies primarily to the nation, but Shpet did not limit the range of collectives to this collective alone.

> Language of the nation, just as is the case for the language of any more or less well defined social formation – a class, a profession, a group united by common work or a handicraft, the language of the yard, the market, and so forth – just like an individual language, is a fact of 'natural' speech with all its cross-national, dialectal, and other characteristics, which enters into the milieux of the general social-historical conditions of a given formation. They define a given form of speech as a 'thing' among things that are subject to material-historical and social-psychological explanation. (1996: 79)

Speaking as a member of a collective

Taken together, Bakhtin and Shpet suggest an approach to language that views it as a subject with dialogic overtones, a form of dynamic energy, rather than as a lifeless object. In what follows I shall focus on how this form of energy derives from the ongoing dialogue, the vibrant 'chain of texts' of a speech community. From this perspective, learning a language at an advanced level involves mastering rules at the level of grammar, to be sure, but it also requires mastery of a generalized collective dialogue. Again, it is at this point that some might be tempted to say that we have left off talking about language and have begun dealing with culture, but as stated at the outset, the intellectual tradition on which I am drawing envisions language as extending into such areas. In order to develop this line of reasoning further, I turn to two illustrations.

Illustration 1: Using expressions that belong to others

In the 2004 U.S. presidential campaign many supporters of George W. Bush accused his opponent John Kerry of 'flip-flopping'. This refrain became so widely used in the Bush camp that his followers took to waving flip-flop sandals at campaign events. Regardless of whether or not one agrees with the charge, during the fall of 2004 the term 'flip-flop' took on a new meaning in American cultural life.

In an attempt to explore what occurred here, I begin by doing something that Bakhtin specifically questioned – turning to a dictionary. As defined by the American Heritage Dictionary[4] 'flip-flop' refers to: '1. A backward somersault or handspring. 2. *Informal* A reversal, as of a stand or position. 3. A backless, often foam rubber sandal'. These definitions are useful in that the second one provides something like the 'literal' meaning at issue and the third one reveals why the act of waving flip-flop sandals could take on its particular significance in the fall campaign.

However, in accordance with Bakhtin's dictum that, 'It is not after all, out of a dictionary that the speaker gets his words!', there is something more to the story. By being used so widely and repeatedly in the 2004 campaign, it is now difficult, at least in the immediate aftermath of the campaign, for a speaker of American English to use the term 'flip-flop' without hearing the dialogic overtones of those who levelled the charge against Kerry. It was nearly impossible in the months following the 2004 presidential campaign to use the term as if it belonged to no one. Instead, using it involved parody – either humorous or bitter – or some other form of double voicedness. Because the term had come to occupy a prominent place in the generalized collective dialogue of contemporary America, speakers of English in this country could no longer use it naively, as if it had no connection to the charges of the 2004 campaign. It is a term that had clearly become 'half someone else's' and hence could not be employed as if it came out of the dictionary.

As a second example of using expressions that belong to others I turn to American political discourse from another era: the discourse surrounding the Watergate scandal during Richard M. Nixon's presidency. In the early stages of this scandal, which ended with Nixon's resignation in 1974, White House press spokesman Ronald L. Ziegler made multiple public statements about the events under investigation by federal authorities and by congress. He initially asserted that the break-in at the Watergate apartment complex that set off the Watergate scandal was a 'third-rate burglary', and for several months he continued to dismiss any media accounts that suggested otherwise.

As events progressed, however, it eventually became clear to others, and eventually to him, that he had not had access to the most damning evidence that was known to insiders at the White House. In the end, information that White House Counsel John Dean revealed in his testimony before a federal grand jury and that other aides were conveying in ongoing investigations led Ziegler to disavow his earlier dismissals and denials. On 17 April 1973, after some particularly embarrassing disclosures had surfaced, he announced to the press that previous White House statements on the issue were 'inoperative'.

If one turns to a dictionary to look up 'inoperative', nothing is mentioned in the definition about Watergate or Ronald Ziegler. Yet for those who lived through the Watergate scandal, the term took on dialogic overtones from Ziegler's unforgettable use of it. At least for a period after the scandal one could not use the term innocently, as if it had not existed 'in other people's mouths, in other people's concrete contexts, serving other people's intentions' (Bakhtin 1981: 294). For example, it would have been nearly impossible for a professor to walk into a classroom and say that earlier comments about the grading policy were inoperative, while keeping a straight face. 'Inoperative' became a term that could only be used in a parodic fashion, given the context of the generalized collective dialogue that existed at the time.

These brief analyses of 'flip-flop' and 'inoperative' touch on several points having to do with generalized collective dialogue. First, they tell us something about the nature of the collectives that are involved. These are collectives defined by a generalized dialogue, not by a linguistic code in some more abstract sense. For instance, it is unlikely that speakers of English in the United Kingdom or India hear the term 'flip-flop' as belonging to the Bush campaign when they used it in the months following the 2004 election.

Second, in addition to being limited in terms of the space across which they apply, the phenomenon of using terms that belong to others is often temporally restricted. For most people, 'inoperative' has now lost the dialogic overtones it had in the immediate aftermath of Watergate. If it still has such overtones, it is only for those who were watching Watergate events closely – perhaps obsessively – in 1973. Many of them may still be unable to use 'inoperative' without hearing Zieglerian overtones in 2005, but today's university students, for instance, are very unlikely to hear them.

Third, the staying power of a generalized collective dialogue may vary. Whereas 'inoperative' may be a term that continues to belong to Ronald Ziegler for the rest of my generation's lifetime, it seems unlikely that we will hear the dialogic overtones of others three decades from now when we ask for a pair of flip-flops in a store.

And finally, while both 'flip-flo[illegible]operative' may be similar in that they reveal ways that terms can b[illegible]ogic overtones', the source of these overtones differs in the [illegible]'flip-flop' are clearly to be found in the Bush campa[illegible] not hear the voice of a single, identifiable indivi[illegible] of us who hear the dialogic overtones in 'inope[illegible]o Ronald Ziegler, an individual, although one wh[illegible] political and cultural debate. In the end, howev[illegible] cases is that the terms' meaning and use are fun[illegible] inclusion in a generalized collective dialogue.

With regard to ad[illegible]on, the point of these examples is that the dialo[illegible] to terms are part of what must be mastered if one is to cla[illegible]dge of a language. While it might be perfectly correct from a gramm[illegible]point of view to use 'flip-flop' or 'inoperative' without recognizing how they belong to others, it is precisely such cases that are likely to be met by the response, 'That is perfectly correct grammatically, but a speaker of our language would never say it that way.' Or, 'You can't really use that word that way.' Such responses point to the importance of the second level of phenomena Bakhtin had in mind under the heading of language, a level concerned with generalized collective dialogue. While some might view such phenomena as belonging to culture, from the perspective outlined here, they belong under the heading of language.

Illustration 2: Responding to others in generalized collective dialogue

A second way that advanced language capacity involves Bakhtin's 'forms of combinations of forms' has to do with the motives behind an utterance. When learning a new language, one is sometimes struck by the fact that the form and overt content of what native speakers say are quite clear, but it is not obvious *why* they seem to feel so compelled to make another comment. In this connection it is often useful to examine how utterances are responses to the voices of others in an ongoing cultural conversation, a conversation that an outsider (i.e., someone who is not an advanced learner of the language) may not hear.

Bakhtin explored this sort of phenomenon under the heading of 'hidden dialogicality'.

> Imagine a dialogue of two persons in which the statements of the second speaker are omitted, but in such a way that the general sense is not at all violated. The second speaker is present invisibly, his words are not there, but deep traces left by these words hav[illegible]termining influence on all the present and visible words of the first spe[illegible]se that this is a conversation, although only one person is speakin[illegible]versation of the most intense kind, for each present, utte[illegible]reacts with its every fiber to the invisible speaker, points [illegible] beyond its own limits, to the unspoken words of [illegible]

[illegible] recognized at the level of local dialogue. [illegible]ock' makes sense only in the context of the [illegible]words that are 'present invisibly' such that the [illegible]spond and react' to them.

[illegible]vant for my present purposes, however, is how hidden di[illegible]rates at the level of Bakhtin's 'language system' that is 'conven[illegible]hin a given collective'. In order to illustrate how this works and what [illegible] implications are for advanced language learning, I turn to a quite different language collective than the ones examined above. Specifically, I shall examine some essays written by young people in the Republic of Georgia in 2003 on an issue that vexes their nation more than any other: breakaway regions that threaten the 'territorial integrity' of Georgia.

The major problem facing Georgia today is Abkhazia, a breakaway region located along the Black Sea in the north-west region of the country. This 'unrecognized republic', as it is called by international observers, was the site of a bitter military conflict in the early 1990s that resulted in brutality, ethnic cleansing (ethnic Georgians were pushed out of the region) and several thousand deaths. Since that time it has been an unrecognized republic, and its leadership has forged close links with Russia (which borders on its north).

Opinions vary as to the forces behind this 'frozen conflict'. Georgians view it as a reflection of Russian machination, including that of oligarchs and the mafia who fear losing the lucrative smuggling and spectacular seaside mansions they have built up over the years. From this perspective, Abkhazia is a natural part of Georgia and should remain so, and Russia should stop its interference and support of the Abkhazians. Russians argue that this region is a volatile area in its 'near abroad' that needs the steadying hand of a major power in light of the incapacity of a weak state like Georgia to deal with it. And ethnic Abkhazians (about 100,000 in the territory) argue that they are a separate nation deserving a separate state – or at least the right to decide whether they want to be part of the Russian Federation.

With this as background, I turn to the essays written by eight young women who are in the American Studies Program at Tbilisi State University. They were asked to write a short essay on the history of Abkhazia and its relationship to Georgia. The essays varied on many points, but on one there was unanimity: that Abkhazia is a natural part of Georgia and has been so for a long time. On this issue students made statements such as: 'Historically, Abkhazia has always been a part of Georgia', 'Abkhazia is a part of my native country named Georgia without which I cannot even imagine my country's existence', 'Abkhazia is one of the oldest and most beautiful areas in Georgia, and it has always been Georgia's inseparable part, indigenous land' and 'Abkhazia is a part of Georgia and can never be the independent state. Abkhazians are in no case an ethnic minority. They are not a different nation and have no basis for establishing an independent state. They have the same alphabet, same religion as we have. So it's not an ethnic minority problem.'

The unanimity and intensity of these statements stand in contrast to what most of the essay writers know exists in the way of legitimate opposing viewpoints. In fact, one of the students provided a quick summary of opposing perspectives in the following terms:

> According to the first approach, Georgians lived in Abkhazia from ancient times and it has always been an integral part of Georgia; Georgians and Abkhazians have always lived brotherly lives. Then the Russian government provoked a conflict hoping to separate Abkhazia from Georgia and expand its territories in the Black Sea region. Another view says that Abkhazia was never a part of Georgia, [that] Georgians came to live on this territory after Abkhazians and they have always been different people in terms of beliefs, culture, and religion. So, Abkhazians decided they had every right to be a separate state and turned for help to Russia, who kindly offered assistance. Both viewpoints are completely opposite except for the last part, pointing to the fact that the third party played a significant role in the conflict. (Student #3)

The real focus of this student's argument emerges in the last line. She went on in her essay to argue that regardless of who was in the region first, the Abkhazians have always lived peacefully with Georgians and it was only with the troublesome meddling of Russian forces that problems began. Moreover, this writer has clear ideas about who is closer to the Abkhaz in cultural terms:

> It wasn't by chance that Russian language became very popular in Abkhazia [i.e., its 'popularity' was encouraged, if not enforced by Russian educational efforts]. It

> should also be noted that Abkhaz people do not have their alphabet. They had to choose between Georgian, Latin and Russian alphabets [and ended up with] the Russian one, although every Abkhazian knows perfectly well that the sounds in the Abkhaz language can best be expressed by Georgian letters. (Student #3)

The most striking general fact about these essays is how adamant the students were in claiming that Abkhazia is part of Georgia. One reading of this is that the students' assertions are simply motivated by what they believe to be true. From this perspective, they made strong statements about this issue for the same reason they would about the fact that Georgia was annexed by the Soviet Union in 1921. To question the truth of such statements would simply fly in the face of what everyone knows and what objective evidence supports.

But all the students – not just the one who mapped out the two approaches – were familiar with historical analyses from legitimate sources that dispute their assertion, so this reading is not very persuasive. Instead, another motive seems to be involved. Namely, the students seemed to be responding – often defensively – to another perspective, or voice in a hidden dialogue. Their essays seem to be part of an 'internally polemical discourse', a category that is related, though distinct from hidden dialogicality in Bakhtin's analysis (1984: 196). According to him, internally polemical discourse involves 'the word with a sideward glance at someone else's hostile word', and it 'cringes in the presence or the anticipation of someone else's word, reply, objection' (1984: 196).

The nature of the hidden dialogicality and internally polemical discourse, along with the defensiveness that grows out of it, may not be immediately obvious to someone who is not part of the generalized collective dialogue in which these students are writing. The obscure nature of the dynamics in this case derives in part from the fact that a *generalized* dialogue is involved. The students were not arguing against a particular individual who had confronted them with an opposing viewpoint, and they were certainly not arguing against someone who was in their immediate speech situation. Instead, they were responding to the voice of a generalized other in their cultural context. The fact that this is a *collective* dialogue stems from the observation that these students were all part of a group that speaks in the same voice. And their essays are part of a *dialogue* in the sense that their texts seem to be organized in response to the perspective of another collective with a different generalized voice.

Recognizing that the dynamics of internally polemical discourse provide the main motive for the students' assertion that Abkhazia provides some insight into their essays, but it does not tell us *which* collective they are responding to. At first glance, it would appear to be Abkhazians, or perhaps their leadership. However, part of the insider knowledge that drives their comments is that they were responding primarily to Russia and its leadership.

Some insight into the generalized collective dialogue that lay behind these students' essays can be gleaned from a comment by Russian President Vladimir Putin, someone who officially personifies the Russian perspective.

Putin made his comment at a meeting in September 2004 after the massacre at the school in Beslan, a city in the North Ossetia province of Russia. In an open-ended discussion with academics and journalists at his residence outside Moscow he stated that Georgia is an 'artificial state'.[5] What he meant by this is that Georgia has no real or natural territorial integrity because Stalin cobbled together various territories into what is now the country simply for temporary political expediency.

Putin made this comment well after the students had written their essays, so they were not responding to it directly, a point that reinforces the generalized nature of this collective dialogue. His comment simply made public a claim that had long been part of an ongoing debate between Russia and Georgia – at least at official levels. The fact that he said this in a state of anger and frustration does not detract from the argument that it is part of such a collective dialogue; in fact, it would appear to strengthen this claim.

Conclusions

The illustrations I have presented touch on only a couple of ways that generalized collective dialogue plays a role in language. I have argued that hearing the dialogic overtones of terms that belong to others and hearing utterances as responses in an ongoing cultural conversation are part of what it means to know a language well. These illustrations suggest that something more than mastering the systems of sound, grammar and meaning must be taken into consideration. Instead of taking language to be an inert code, the ideas of figures such as Bakhtin and Shpet suggest that it must be viewed as a dynamic form of dialogic energy.

As noted in my introduction, such claims may be resisted by some readers because they assume that generalized collective dialogue falls outside the legitimate realm of language. From their perspective, what I am discussing may be of interest as a cultural phenomenon, but it does not belong under the heading of language *per se*. To be sure, generalized collective dialogue goes beyond the realm that Saussure considered the legitimate province of his analysis, and it also would not qualify as appropriate subject matter for many contemporary approaches in linguistics. However, it does qualify as a legitimate object of analysis for figures like Bakhtin and Shpet.

In the course of making this argument I have distinguished between generalized collective dialogue, on the one hand, and localized dialogue, or what Bakhtin termed the 'primordial dialogue of discourse', on the other. Localized dialogue is widely recognized as part of what the study of language must account for, but generalized collective dialogue is not. In trying to account for advanced language acquisition, however, it is an issue that will eventually have to be taken into account.

Engaging in discussions of advanced language acquisition inevitably introduces new questions and perspectives into linguistic studies. No one who has ever tried to learn a second language can doubt the importance – and difficulty – of mastering systems of sound and grammar. However, most such learners at some point encounter difficulties that go well beyond these issues but still seem to be part of language study. As noted above, these difficulties are indexed by responses to an utterance such as, 'That is perfectly grammatical, but we would never say it that way.'

Such cases suggest a need to invoke an expanded notion of language, one that takes it to be *energeia* as recognized by Humboldt and Shpet. Specifically, they suggest the need to recognize language as a dynamic force that derives from generalized collective dialogues. To use a language at a high level of expertise, then, one must recognize how expressions are situated in, and carry the force of this form of dialogue. From this perspective it is surely correct, as Bakhtin states, that words do not get their meaning out of a dictionary alone.

In the end, this does not answer the question I laid out at the beginning of this paper, the question of where language stops and culture begins. However, the line of reasoning I have outlined does suggest that analyses of language that will be relevant for advanced language learning might need to account for phenomena that we often do not include under this heading. It may very well be that the notion of generalized collective dialogue that I have introduced will need to be expanded or elaborated to meet the demands of the task I have outlined, but following Bakhtin and Shpet, it seems to be a good place to start.

Notes

1 The writing of this chapter was assisted by a grant from the Spencer Foundation. The statements made and the views expressed are solely the responsibility of the author.
2 This is a term in Russian that can also be translated as 'utterance', a term that I shall sometimes employ in what follows.
3 The term 'slovo' in Russian is often translated into English as 'word', but for figures such as Bakhtin and Shpet it clearly means something broader than a decontextualized vocabulary term or lexical item. In fact, some translators of Bakhtin have used the term 'discourse' for this term.
4 *The American Heritage Dictionary*. Fourth edition. Boston: Houghton Mifflin Company, 2001.
5 Nikolai Zlobin in Caucasus International Forum Roundtable, *Caucasus Context*, Issue 3, 2005.

References

Bakhtin, M. M. (1981) *The Dialogic Imagination: Four Essays by M. M. Bakhtin*. Austin, TX: University of Texas Press. (Edited by M. Holquist; translated by C. Emerson and M. Holquist.)

Bakhtin, M. M. (1984) *Problems of Dostoyevsky's Poetics*. Minneapolis, MN: University of Minnesota Press. (Edited and translated by C. Emerson.)

Bakhtin, M. M. (1986a) *Speech Genres and Other Late Essays*. Austin, TX: University of Texas Press. (Edited by Caryl Emerson and Michael Holquist; translated by Vern W. McGee.)

Bakhtin, M. M. (1986b) 'The problem of speech genres', in M. M. Bakhtin, *Speech Genres and Other Late Essays*, pp. 60–102.

Bakhtin, M. M. (1986c) 'The problem of the text in linguistics, philology and the human sciences: an experiment in philosophical analysis', in M. M. Bakhtin, *Speech Genres and Other Late Essays*, pp. 103–31.

Holquist, M. (1986) 'Introduction', in M. M. Bakhtin, *Speech Genres and Other Late Essays*, pp. ix–xxiii.

Shpet, G. G. (1996) 'Vnutrennyaya forma slova', in T. D. Martsinkovskaya (ed.), *Psikhologiya Sotsial'nogo bytiya. Izbrannye Sotsiologicheskie trudy.* Moscow: Institut prakticheskoi psikhologii, Voronezh: MODEK, pp. 49–260. (Originally published in 1927)

Voloshinov, V. N. (1973) *Marxism and the Philosophy of Language.* Translated by L. Matejka and I. R. Titunik. New York: Seminar Press.

Vygotsky, L. S. (1987) *The Collected Works of L.S. Vygotsky. Volume 1. Problems of General Psychology.* Including the Volume *Thinking and Speech.* New York: Plenum. (Edited and translated by N. Minick.)

3 Re(de)fining language proficiency in light of the concept of 'languaculture'*

James P. Lantolf

Introduction

Definitions of advanced speaking proficiency, in particular the Interagency Language Roundtable (ILR) guidelines, generally seen as the gold standard in the United States, assume that attaining high levels of language proficiency requires facility with cultural knowledge. Thus, the descriptor for Level 2+ speaking ability states that 'the individual may miss cultural and local references', while the descriptor for advanced proficiency, or what is often referred to as 'distinguished proficiency' – Level 4+ proficiency – asserts that 'the individual organizes discourse well, employing [*sic*] functional rhetorical speech devices, native cultural references and understanding'. However, even at the distinguished level 'the individual would not necessarily be perceived as culturally native' and therefore 'occasional weaknesses in idioms, colloquialisms, pronunciation, cultural reference' are anticipated and the speaker is not expected 'to interact in a totally native manner' (ILR).

The interesting aspect of the descriptors, however, is not the inclusion of cultural facility at upper levels of proficiency; rather it is the dichotomy between language and culture that pervades the guidelines and much of the thinking about proficiency that they have inspired. For example, in the recent white paper published by the Center for the Advanced Study of Language (CASL 2005), *A Call to Action for National Foreign Language Capabilities*, I counted no fewer than forty-one dichotomous mentions of language and culture, as illustrated by the following excerpts:

> 'critical need to take action to improve the foreign language and cultural capabilities of the Nation' (ii).
>
> 'Foreign language education in primary schools, secondary schools, and post-secondary institutions should ensure continuity of language and cultural instruction through the advanced levels' (11).
>
> 'Government-sponsored research and evaluation programs should be implemented to help identify and support innovative academic approaches to teaching,

study abroad, immersion, and other traditional methods used to acquire language and cultural skills' (11).

'Federal, state, and local governments, as well as officials who implement language assistance under Title VI of the Civil Rights Act of 1964, need individuals with bilingual and bicultural capabilities . . .' (7).

In the *Executive Summary* of ACTFL's document entitled *Standards for Foreign Language Learning. Preparing for the 21st Century*, culture constitutes one of the five C's of foreign language education. Yet, as the following statement shows, while the document brings language and culture into a closer nexus than, perhaps, they have been in the majority of foreign language programs to date, they are still seen as separate entities:

> Through the study of other languages, students gain a knowledge and understanding of the *cultures* (italics in original) that use that language and, in fact, cannot truly master the language until they have also mastered the cultural contexts in which the language occurs. (Online document)

The intent of the present chapter is twofold: to argue against the language/culture dichotomy as an unfortunate consequence of the attempt to construct linguistics as a science and to argue for a unified approach to language/culture grounded in Agar's (1994) concept of 'languaculture'. In the remainder of the chapter I will first briefly address the consequences of the bricolage between language and culture created by Saussure and reinforced by Bloomfield in their attempts to convert linguistics from a tool for conducting anthropological field work into an independent scientific discipline. Next, I will discuss Agar's concept of 'languaculture' in an attempt to reunite what Saussure and Bloomfield tore asunder. I will then consider some L2 research that illustrates the significance of this concept and what it may mean for how we conceive of advanced proficiency and for how we design pedagogical programs that promote its development.

The language/culture divide

Fauconnier and Turner (2002: 3) remark that 'we live in an age of the triumph of form' in which knowledge has been reduced to 'a matter of essential formal structures and their transformation'. Arguably, the two scholars most responsible for the triumph of form in linguistics, and with it the ostensible conversion of linguistics into a scientific discipline, are Ferdinand de Saussure and Leonard Bloomfield (Agar 1994). Saussure recognized that science, in particular physics, was in the business of studying the structure of the material universe, which, importantly, exists independently of the scientists who observe and analyse it. To attain his goal of establishing linguistics as a scientific discipline, Saussure claimed that language and the world of objects studied by the other sciences (e.g., physics, biology, chemistry) 'belong to the same ontological order' (Crowley 1996: 18). Once language was conceived of as a 'thing to be found in the world of other things' it attained the 'privileged status of scientific object' and could therefore be 'open to the methods of objective scientific study' (*ibid.* 18).

Saussure succeeded in constructing language as a scientific object by first distinguishing language (*langue*) from speech (*langage*) and subsequently arguing that because speech is 'many-sided and heterogeneous' and belongs simultaneously to 'the individual and to society' it cannot be 'put into any category of human facts' (1959: 9). As Agar (1994: 37) succinctly puts it, 'speech is a mess'. Saussure asserted that linguistics could not lay unique claim to the study of speech, in large part because its research method for such study was 'faulty'; therefore other disciplines, including psychology, anthropology, normative grammar and philology, could appropriately declare speech as their own object of study (1959: 9). Language, on the other hand, according to Saussure, 'is a self-contained whole and a principle of classification' that 'seems to lend itself to independent definition' as '*the norm of all other manifestations of speech*' (italics in original). It is, in Agar's (1994: 37) words, 'pure, clean, a steel skyscraper arising above the chaos of the streets'. For Saussure, then, language, as the 'inventory of symbols with a system that ties them together' (Agar 1994: 37) was indeed the unique and proper object of study for the newly created science of linguistics. The system of linguistic signs that comprises language, according to Saussure (1959: 15), is no less concrete than speaking and constitutes a reality with its seat in the brain.

Raising the status of language study from mere tool in the fieldworker's kit to that of a scientific discipline in its own right was not without cost. In order to reify language into an object of scientific study Saussure was forced to work with speech, as the observable manifestation of the norm he was after. This required, however, that speech be sanitized, which meant abstracting it from human activity and 'the realm of history' (Crowley 1996: 18) and with that the realm of culture as well. According to Saussure, 'language presupposes the exclusion of everything that is outside its organism or system – in a word, of everything known as "external linguistics"' (1959: 22). External linguistics addresses those aspects of language 'that we think of when we begin the study of speech' including features emanating from the influence of culture on language (1959: 4). In essence Saussure drew a circle around language (Agar 1994: 41) and proposed that inside-the-circle language, the proper and exclusive domain of linguistic science, was restricted to the study of grammar and dictionary; that is, the study of form and 'core meaning' or 'the part of meaning that can be characterized formally and truth-conditionally', and therefore considered as 'the only important and fundamental part of meaning' (Fauconnier and Turner 2002: 15). So-called non-core meaning (e.g., figurative meaning) was marginalized, if not placed outside-of-the-circle. For Saussure outside-of-the-circle features of language were not uninteresting and unworthy of study, but he insisted that their study was not required to 'understand the internal linguistic organism' (1959: 22).

Bloomfield drew the Saussurean circle even tighter and to a large extent even dictionary meaning was expunged from inside-the-circle. Linguistics thus became 'the study of the sound system and the grammar' (Agar 1994: 55). According to Bloomfield (1933: 140), 'the statement of meanings is therefore the weak point in language-study' and to define the meaning of a form linguists must appeal 'to students of other science or to common knowledge' (1933: 145). Thus, the linguist can only define *went* as the past of *go* 'if the meanings of the English past tense and of the word *go* are defined'; similarly, if the meanings of *male* and *female* are 'defined *for* (italics added) the linguist', the linguist 'can assure us that they represent the difference between *he*: *she*, *lion*: *lioness*, *gander*: *goose*, *ram*: *ewe*' (1933: 146). In this way, Bloomfield kept 'the circle's edges clear and intact', and thus allowed linguists 'to leave out the study of culture and go their own way' (Agar 1994: 39), which they did.

While linguistics was able to tease out important insights about inside-the-circle language, it failed to uncover equally important insights about outside-the-circle language. Perhaps more significantly, however, the circle compelled us to think of language as only what exists inside its boundaries. What is outside-of-the-circle is something other than language, for example, culture. We are then left with the language/culture dichotomy mentioned at the outset of our discussion. Little wonder that language learning, teaching and assessment have been understood as something apart from the learning, teaching and assessment of culture. One phenomenon resides inside the circle, and is therefore of primary concern, and the other resides outside the circle and, with some exceptions within present-day applied linguistics (e.g., Kramsch 2004), is of secondary importance, at least until advanced levels of proficiency are considered; yet, even here, as I have pointed out, language is seen as separate from culture.

Ironically, although Saussure's circle cut language off from cultural meaning, at the same time, he set the stage for eventual erasure of the circle (Agar 1994: 47). Linguistics, for Saussure, was established as 'a part of the general science of semiology' – '*a science that studies the life of signs within society*' (italics in original) and which, when fully developed, would form 'part of social psychology and consequently of general psychology' (Saussure 1959: 16). The linguist's task is to discover 'what makes language a special system within the mass of semiological data' (*ibid.*). To achieve this end and to avoid 'going around in circles' (how prophetic this statement would become!), language, in Saussure's view, must be studied 'in itself' rather than 'in connection with something else, from other viewpoints' (*ibid.*). On the other hand, he also argued that the only way to discover the true nature of language is 'to learn what it has in common with all other semiological systems' (1959: 17).

No doubt Bloomfield's tightening of the circle contributed significantly to the failure of Saussure's original semiological project, and, as a result, linguistics, particularly in North America, isolated itself from the study of signs in other domains (e.g., film, literature, fashion, law). Chomsky's mathematical linguistics pulled the study of language even further away from anthropology and 'the messy world of bumbling speakers and hearers hammering out reality and getting through the day with language' (Agar 1994: 114).

Agar erases the Saussurean-Bloomfieldian circle, and in so doing reunites speakers with language and thus makes language whole again. The concept he uses to capture this reunification is '*languaculture*'. According to Agar (1994: 60), languaculture is intended to establish the '*necessary* (italics in original) tie between language and culture' so that whenever we hear language or culture used as individual terms we will 'wonder about the missing half'.

Languaculture: a brief exegesis

A very important consequence of Agar's move to reunify language and culture is that meaning is brought whole cloth back into the picture. It is the 'thread that tied language and culture together' (Agar 1994: 71). The concept of languaculture establishes a much more complex and a far richer relationship between signifier and signified than is the case in inside-the-circle linguistics. It entails 'basic assumptions about what is *significant* (italics in original) in the world at large' (1994: 71). These basic assumptions are the theory of reality that comprises a culture. For Boas, study of the signifier, language, was the means to study the theory of reality of a community – its culture; for Whorf, however, 'studying language and studying culture *were the same thing*' (italics in original) (Agar 1994: 71). Linguistics, then, for Whorf is 'essentially the quest for MEANING' that illuminates 'the thick darkness of language, and thereby of much of the thought, the culture, and the outlook upon life of a given community' (Whorf 1956: 70).[1]

Whorf's contention was that languaculture 'shapes *consciousness* (italics in original), shapes ways of seeing and acting, ways of thinking and feeling' (Agar 1994: 71). This is because, for Whorf, languaculture comprises not just forms but 'meaningful forms' (Lucy and Wertsch 1987: 73). These forms organize reality by classifying things and events together 'which are in many ways quite different' and thus suggest to their users 'associations which are not *necessarily* (italics in original) entailed by experience' (Lucy and Wertsch 1987: 73). Given that different languacultures utilize different classificatory systems, participants in different systems will have different experiences of reality, but will assume that their way of talking and thinking about reality is natural and obvious for all to see. Thus, English classifies time as something tangible and equivalent to an object that has substance such as 'a book'. Users of this languaculture pluralize and quantify time as they do books (e.g., books, days, five books, five days) and they use it as an argument of possessive verbs such as 'have' (e.g., I have five books, I have five days left before the exam). In Hopi, on the other hand, pluralization often results in the assignment of animateness to what is, in the singular, an inanimate object. Thus, a singular inanimate noun, such as *ʔo:*'mâw 'cloud', is marked as animate when it is pluralized (Whorf 1956: 82).

Following Sapir's proposal 'that each language shapes the conceptual world of its speakers' (Lucy and Wertsch 1987: 73), Whorf argued that thought becomes 'bound to a cultural perspective' through the enculturation of a particular languaculture (Lucy and Wertsch 1987: 80).[2] According to Agar, however, this is not an argument for linguistic determinism, but an argument for linguistic relativity. Thus, while languaculture does not imprison its users behind impenetrable perceptual walls, it does lay down 'comfortable ruts of perception' that people generally stay in because they know them and 'function quickly and efficiently within them' (Agar 1994: 71). This does not, however, preclude moving from one rut to another; but, as I will discuss later, to do so is not necessarily easy (for similar observations, see the contributions by Matthiessen, this volume, von Stutterheim and Carroll 2006, and Carroll and Lambert 2006).

Communication and cognition

Although Whorf argued that languaculture shapes consciousness, he did not develop an explicit theory of how this happens (Lucy and Wertsch 1987: 73). He merely assumed that languaculture was something people have rather than something that happens to them (Agar 1994: 72). Consequently, he was left to infer the cognitive effects on real people of the lexicogrammatical differences he uncovered in his cross-cultural research. Thus, while Whorf brought language and culture back together, he still left concrete people engaged in goal-directed human activity out of the picture.

As it turns out, at roughly the same time that Whorf was carrying out his research on linguistic relativity, L. S. Vygotsky was developing an explicit theory of mind in which organized semiotic systems, language being the most pervasive of these, mediate not only social communication but mental functioning as well.[3] Through participation in semiotically mediated interaction with other members of a culture, mental processes such as 'categorization and dialogic reasoning emerge' (Lucy and Wertsch 1987: 71). Thus, Vygotsky held what Clark (1998) refers to as a 'supra-communicative' conception of language (see also Wertsch's formulation of 'linguistic imperialism', this volume). On this view, there is an organic relationship between communication and thought whereby social communication comes to serve as a tool for mediating thinking. This is in sharp contrast to the traditional view in mainstream linguistics and psychology, in which language and thought are completely independent phenomena and where language serves only to transmit thought but is in no way implicated in its formation (for extensive discussion of the relation between language and thought, taking a systemic-functional linguistic perspective, see Matthiessen, this volume).

According to Lucy and Wertsch (1987), although Whorf and Vygotsky both made concepts the focal point of their thinking, they differed in four important respects. First, Whorf was more interested in the meanings carried by grammatical concepts, whereas Vygotsky was more interested in the meanings carried by words. Second, Whorf considered grammatically organized concepts independently from actual uses of these concepts in concrete activity, while Vygotsky was concerned with how concepts developed as children grew into adults and how concepts functioned to mediate thinking as individuals carried out specific psychological activities. Third, Whorf foregrounded habitual thought that people use in their everyday activity over more formal thought used by the scientific community, while Vygotsky focused on the development of habitual thinking (for Vygotsky, spontaneous thinking) into scientific thinking that occurs in formal educational settings. Fourth, Whorf's research methodology was grounded in comparative analysis of different languacultures and Vygotsky's approach to research, based on Marxist historical-dialectical philosophy, was genetic to the extent that he focused on how thinking in individuals developed over time as they appropriated the semiotic systems of their languaculture.

In my view, Lucy and Wertsch's comparison of the two great scholars requires some modification. While it is the case that Vygotsky was intently interested in the development of word meaning in the ontogenesis of children, he was also interested in conceptual meanings carried by lexicogrammar. Furthermore, although he did not carry out primary comparative research of languacultures, he was clearly interested in the relevance of such work for a theoretical model of mind. Both of these points are evidenced in the monograph he co-authored with A. R. Luria, *Ape, Primitive Man, and Child. Essays in the History of Behavior*, which appeared in English in 1992. In this work, the authors discuss the role that semiotic mediation plays in three genetic domains: phylogenesis, in which human thinking is compared to higher primates; ontogenesis, in which the development of thinking in children over time is examined; and the sociocultural domain, in which the authors review the comparative research of scholars such as Levy-Bruhl, Thurnwald and Wertheimer, who analysed the language of indigenous communities and proffered arguments for what these systems meant for the mental life of their users.

A salient and intriguing difference between Vygotsky and Whorf in this domain and, unfortunately, one which I cannot pursue here, is that Whorf did not consider one form of culturally contextualized thought as superior to another and therefore 'minimized the significant historical evolution of the uses of language in thought' (Lucy and Wertsch 1987: 84). He argued that any analysis of reality was necessarily provisional and that proclamations to finality were illusory, including those of Western scientific thinking (Lucy and Wertsch 1987: 85). Vygotsky, on the other hand, stressed the historical development of the ways in which language is implicated in thinking. Vygotsky proposed that different communities use 'words in quite different ways. Words can be put to different functional uses. The mental operations performed with a word will also depend on how it is used' (Luria and Vygotsky 1992: 69). For example, the emergence of literacy in a community has profound consequences for what kinds of concepts develop and for how they are used to mediate thinking (see Luria 1976 and Olson 1994). He proposed that in some cases, such as in indigenous societies, words are used to individuate and to capture 'a photographic description of an event with the finest details' (Luria and Vygotsky 1992: 62), while in others, such as in modern technological societies, and in particular within the scientific communities that emerged in these societies, words show a strong tendency to generalize and categorize. For Vygotsky – and this opened a controversy that has yet to be adequately dealt with – the latter use of language was historically more developed than the former.[4]

Languaculture proficiency in a second language

Once the Saussurean–Bloomfieldian circle around language is erased and we begin to think in terms of languaculture rather than language *and* culture, meaning becomes far more prominent than it is inside the circle. This has consequences for how we conceptualize L2 learning and L2 proficiency. Inside the circle, meaning, as I have pointed out, is for the most part referential and in terms of L2 learning amounts to knowing that, for example, the Spanish word for 'fork' is *tenedor*. Outside of the circle, the domain of languaculture, meaning becomes much more interesting and complex because it entails knowledge of different concepts and how these are encoded in such features as conceptual metaphors, lexical networks, lexicogrammatical structures, schemas and the like that represent different ways of organizing the world and our experiences in it (Kecskes and Papp 2000; Shore 1996).

Agar (1994: 100) points out that differences in conceptual frameworks are most often made visible at *rich points*, or what he eloquently describes as 'Whorfian cliffs', where languacultures come not just into contact – contact that is often conflictual – but where the person has the opportunity to develop new ways of perceiving, talking and thinking about reality. In the next section, I will briefly discuss some recent research that focuses on conceptual rich points encountered by L2 learners as they attempted to speak about a series of motion events through their new language. As we will see, the learners were generally unable to take advantage of the rich points to develop new ways of dealing with the reality motion offered by their L2. My argument is that language proficiency in general, and advanced language proficiency in particular, entails the appropriation of conceptual meanings that emerge at rich points.

The concept of motion and the relevance of gesture

Slobin (1996, 2003) has recently proposed a 'thinking for speaking' (henceforth, TFS) hypothesis in which thinking takes on a particular quality as experiences are filtered through language into verbalized events. The hypothesis suggests that speaking through a particular language (in our terms, languaculture) not only influences how people talk about events but, more importantly, how they experience those events 'they are likely to talk about later' (Slobin 2003: 179). This Slobin calls the 'anticipatory effects' of language, which arise during *experience time* when 'prelinguistic or nonlinguistic coding' takes place as the person attends 'to those event dimensions that are relevant for linguistic coding' at *speaking time* when the person attends to, and accesses, 'the linguistically codable dimensions' of the event

(p. 179). Slobin's hypothesis is commensurate with Vygotsky's view that speaking completes the thinking process.

Among the topics addressed in Slobin's research is how speakers of different languages deploy TFS to describe motion events depicted in short picture narratives. This work is informed by Talmy's research on the typology of motion verbs. Talmy (2000: 26) categorizes motion events according to six criteria: **Figure** – an object moving or located with respect to another object (ground); **Ground** – a reference object in relation to which the figure moves; **Path** – trajectory or site occupied by the figure; **Motion** – changes of locatedness in the event; **Manner** – the particular way the motion is performed; **Cause** – the efficient origin of a change in motion or location (see also the treatment by Langacker 2006).

Languages can be grouped according to how they express the path of motion events. Satellite-framed, or S-languages (Slobin's convention), such as English, Dutch and German, encode path in particles or adverbs rather than in verbs, while Verb-framed, or V-languages, such as Spanish, Turkish and Korean, conflate path with the motion verb itself. In English, for example, the verb 'climb' does not indicate whether the path trajectory is upward or downward. To indicate path, speakers must use directional markers as in 'Climb up the ladder', 'Climb down the ladder', 'Climb under the fence', 'Climb through the fence', 'Climb out from under the bench'. Spanish, on the other hand, a V-language, has separate verbs to indicate path of motion, as in *subir* 'to ascend' and *bajar* 'to descend.'

While it is essential to indicate path of motion (without path, movement of a figure against a ground, there is no motion), the manner in which a figure moves against the ground is optional. That is, S-languages generally prefer to conflate manner with verbs, while V-languages rarely conflate manner with verbs and either forego the expression of manner altogether or encode it in separate lexical forms, or, as we will see shortly, in gesture. For example, in English 'Tarzan swings through the jungle', and in Spanish *Tarzan salta (de liana a liana) por la selva,* 'Tarzan jumps (from vine to vine) through the jungle,' where the phrase in parenthesis is optional.

The English lexicon is saturated with verbs that conflate motion and manner (e.g., trudge, skip, hop, slid, sidle, shinny, scamper, sweep, leapfrog and countless others) (Slobin 2003: 163).[5] Although languages such as Spanish have manner verbs (e.g, *tambalearse* 'to tumble', *trepar* 'to climb', *agitar* 'to agitate'), compared to English, such verbs are relatively few in number, and they occur with considerably less frequency in everyday speech than they do in English. Thus, English-speaking children are immersed in communication that strongly draws their attention to the 'fine-grained' aspects of motion events. While Spanish-speaking children may have their attention drawn to manner, it will most often be achieved in a much less graphic way (Slobin 2003: 164). Consequently, English speakers are more likely than their Spanish counterparts to develop a 'rich mental imagery of manner of motion'; and therefore 'manner of motion will be salient in memory of events and in verbal accounts of events' (*ibid.*).

David McNeill (1992 and 2000) and his colleagues (e.g., McNeill and Duncan 2000) have shown that, although speakers may not encode manner of motion verbally, they may opt to do so through gestures. McNeill's research expands Slobin's TFS framework and argues that the interaction between gesture and speech during communicative activity presents a more robust picture of how speakers construct thinking for speaking than does analysis of verbal performance alone. To capture this notion, and drawing his inspiration from the writings of Vygotsky on inner speech, McNeill (2000) proposes the concept of *growth point*, a unit of thinking for speaking, or perhaps more appropriately, for communicating, that fuses into a single meaning system 'two distinct semiotic architectures', one verbal and one imagistic. Each of these contributes 'unique semiotic properties' to the growth point and therefore to the thinking process (McNeill and Duncan 2000: 144). Paraphrasing Vygotsky, McNeill and Duncan (2000: 155) suggest that gestures are '*material carriers* of thinking' and therefore provide 'an enhanced window into mental processes' (2000: 144).

When gestures are brought into the picture, the analysis of motion events becomes even more interesting. V-languages and S-languages synchronize speech and gestures in markedly different ways. English speakers, for example, coordinate manner gestures with manner verbs if the primary focus of their attention is on the manner rather than the path of a motion event. If English speakers opt to defocus manner they are still likely to use a manner verb but will forego a manner gesture, and they rarely if ever use manner gestures in the absence of conflated manner verbs (McNeill and Duncan 2000). Spanish speakers, on the other hand, coordinate path gestures with conflated path verbs or with ground NPs. According to McNeill and Duncan (2000: 152), however, signalling manner in a V-language is a challenge. Spanish, unlike English, sanctions use of manner gestures in the absence of a conflated manner verb; moreover, and also unlike English, manner gestures can synchronize with path verbs and ground NPs; and they can be omitted entirely even in cases where manner of motion is 'potentially significant'.

In example (1) below, an English L1 speaker is renarrating a segment from a Tweety Bird cartoon, where Tweety drops a bowling ball down a drainpipe as Sylvester Cat is climbing up through the inside. The speaker focuses on the manner of the character's motion as the bowling ball pushes him out the bottom of the pipe.

(1) (but it **rolls**) him out
MANNER = Hand wiggles: manner information
(McNeill and Duncan 2000: 150)

The crucial feature of the gesture, its *stroke* (that portion involving hand movement synchronized with speech) is indicated by the bolded word in brackets. In this case, manner is doubly marked in speech and in gesture, an indication, according to McNeill's analysis, that manner is in focus.

In (2) a different speaker describes the same event but synchronizes the stroke of the gesture on the satellite 'down' and sustains it (indicated by the double underline) throughout the production of the ground NP 'the drain spout'. In this case, the speaker's attention is on the path of Sylvester's motion rather than on its manner, even though a conflated manner verb is used.

(2) (and he rolls . . . **down** the drain spout)
PATH = Hand plunges straight down: path information only
(2000: 150)

Thus, while both speakers use the same manner verb, they do not think about the event in the same way. Their gestures betray different growth points and therefore we can conclude that they each process the same event in different ways.

In (3) a Spanish speaker, narrating the same scene, uses neither a manner nor a path gesture and instead focuses on the shape of the ground NP, the drainpipe, which is encoded in gesture only.

(3) e entonces busca la ma[ner**a** (**silent pause**)]
'and so he looks for the way'
GROUND GESTURE = depicts the shape of the pipe
(2000: 153)

In (4) the same speaker indicates Sylvester's path both verbally and in gesture as the cat goes up through the drainpipe. The speaker holds the shape of the drainpipe initiated in (3) and moves his hands upward, thus indicating Sylvester's upward motion. Simultaneously, he marks the manner of the cat's motion by rocking his hands back and forth. The verb *meterse* 'to put oneself into' marks path only. The result is what McNeill and Duncan (2000: 151) call a 'manner fog', in which manner is marked only through gesture that spreads (as fog) over the entire utterance. Path, on the other hand, is indicated both in speech and gesture.

(4) [de e**ntra**] [r / / **se met**] [**e por el**] [de**sague** / /] [sí?]
'to enter REFL goes-into through the drainpipe . . . yes?'
PATH + MANNER = *Both hands rock and rise simultaneously: manner and path (left hand only through 'mete'). Right hand continues to rise with rocking motion*
(2000: 151)

Gesture in L2 performance

To my knowledge only five studies on gesture within the TFS framework have appeared so far in the L2 research literature. Four of these have addressed motion verbs: Stam (2001), Özyürek (2002), Kellerman and van Hoof (2003), and Negueruela *et al.* (2004), and one, Gullberg (2005), focuses on placement verbs. Space does not permit a detailed analysis of each study; I will therefore summarize four of the five and will limit consideration of specific examples to the study conducted by Negueruela and his colleagues.[6] In all the studies, the relevant question is whether speakers of an S-language or a V-language are able to master the speech/gesture synchronization patterns of a typologically different language or whether they continue to use the patterns of their L1 when talking about the concept of motion. In essence, the question goes to whether or not L2 learners are able to appropriate new conceptual meanings to mediate their TFS in their new languaculture.

By and large the studies confirm that even in the case of very advanced speakers, it is difficult to move out of the rut established by one languaculture and into a new rut laid down by another. This is captured nicely in the following comment from Özyürek's study:

> the verb and satellite construction used dominantly [sic] by native speakers of English is a hard construction to master for Turkish speakers in L2 and needs years of practice in a country where the L2 is spoken. That is, typologically distinct and different constructions across languages are hard to learn in L2. (Özyürek 2002: 508)

In her study, Özyürek analysed the performance of beginning, intermediate and advanced Turkish L1 speakers of L2 English. She reports that neither of the first two groups, who had studied English in Turkish schools for several years, manifested any evidence of a shift toward English patterns. They continued to follow the Turkish pattern of marking path and manner on separate path and manner verbs in the same utterance.[7] Özyürek also reports, however, that her advanced L2 speakers, who had resided in the United States for about ten years before returning to Turkey to take up teaching positions at a Turkish university, occasionally used English manner verbs (e.g., 'roll') accompanied by gestures that conflated manner and path, a possible English pattern, if the stroke of the gesture synchronizes with the manner verb. Unfortunately, Özyürek does not provide evidence with regard to this important point.

Stam's (2001) study of Spanish L1 learners of English L2 found no evidence of shifts in how the L2 speakers encoded manner in their new language, although she does report some evidence of speakers appropriately marking path through satellite expressions and synchronized path gestures in English. Kellerman and van Hoof (2003) report the odd, and unexplained, finding that L1 speakers of Dutch, an S-language, appeared to use Spanish-like patterns, which synchronized gestures with path verbs rather than satellites when speaking L2 English. The study makes no mention of a shift with regard to the crucial manner of motion.

Finally, Gullberg (2005) presents the findings of a very interesting study on the description of placement events, which, while not true motion events as defined here, comprise a closely related category. Placement events are described in S-languages through a variety of different verbs, such as 'Helen puts/places/lays the bowl on the table', thus allowing for optional focus on the object that is placed rather than its landing site. In V-languages, on the other hand, the inventory of placement verbs is not as robust and the tendency is to focus on the landing site, as in Spanish *Elena mete la olla en la mesa* 'Elena puts the bowl on the table'. Gullberg points out that the gestures that co-occur with placement verbs across the two typological categories also differ. In S-languages, because focus is often on the object, speakers prefer to hold their hands in the shape of the object while moving them toward the envisioned landing site (e.g., both hands, fingers rounded, and palms pointing toward each other iconically representing a bowl). In V-languages, speakers generally forgo representation of the object manually and instead use their hands to point to the landing site. In her study, Gullberg reports that L1 Dutch speakers of L2 French used the appropriate French verb (e.g., *mettre*) when describing a series of placement events, but they continued to use their L1 gesture pattern of shaping their hands to resemble the object placed (e.g., bowl) instead of the French preference of pointing to the landing site. Thus, the Dutch speakers continued to conceptually function as they would in their L1, although linguistically they were able to select an appropriate French verb.

This brings us to the final study, and one which I will describe in a bit more detail than the others. Unlike previous studies, Negueruela *et al.* (2004) included advanced L2 speakers (L1 English) of a V-language, Spanish, and of an S-language, English (L1 Spanish) in research on motion events. In their study all the speakers were graduate students at a U.S. university and all had spent time living in a country where the L2 was spoken. The stimulus in their study was Mayer's (1979) well-known picture story, *Frog Goes to Dinner*, also used in Slobin's research. A typical pattern produced by the L2 English speakers is illustrated in (5), where the speaker narrates a restaurant scene where the frog suddenly leaps out of the old woman's salad and moves toward her face:

(5) the [**frog ap**pears] . . . from inside the salad
MANNER = both hands coming up toward the speaker's face
(Negueruela *et al.* 2004: 136)

The speaker uses the English cognate of the Spanish verb *aparecer*, which would be a common way of depicting the frog's motion in Spanish, and highlights through gesture the frog's path rather than the manner of its motion. One of the L1 Spanish speakers produced essentially the same utterance in Spanish, but without a gesture, to describe the same event: *Le aparece la rana* 'To her appears the frog'.

When the L1 Spanish speakers marked manner in English, they preferred to do so in gesture only and often the gesture was conflated with a path verb. In (6) the speaker narrates a scene where various eating utensils 'fly off' (as described from an English L1 perspective) the dining table as a result of the frog's movements:

(6) and [the **cup**, the **plate**, the **fork** are all **falling** off the table]
PATH + MANNER = four consecutive strokes with both hands, palms facing each other, vigorously moving upward (last stroke more pronounced) (Negueruela *et al.* 2004: 136)

The speaker brings manner into focus as indicated by the vigour of her hand movements conflated with the path gesture showing motion against the ground NP 'table'. This is not a typical English pattern for highlighting manner, since the spoken component does not contain a fine-grained manner verb (e.g., 'fly off'). Moreover, in the absence of upward hand movements indicating path, one could easily assume that the utensils were simply falling to the floor in a downward trajectory, which they were not.

The general impression an English L1 listener might construct on the basis of the L2 narratives produced by the Spanish speakers is that the story was lacking in 'real' action. This is no doubt due in large part to an absence of complex manner verbs – something L1 English users expect to find in such stories. Indeed, this expectancy presented a problem for the English speakers when narrating the story in L2 Spanish. This is illustrated compellingly in (7), where an L2 Spanish speaker attempts to describe a scene where the waiter is carrying a plate of salad in which the frog has hidden. As the waiter walks through the restaurant, the salad begins to move because of the frog's motion.

(7) la ensalada [**está** . . . como en medio aire]
'the salad [is . . . like in mid-air]'
MANNER = hand shaking palm down
(Negueruela *et al.* 2004: 134)

The speaker attempts what appears to be a progressive construction comprising the auxiliary *estar* 'to be' and a verb in the progressive form. She pauses, however, which, according to Negueruela *et al.*, signals a search for a verb that conflates motion and manner. Plausibly, the speaker had encountered a languaculture rich point, where her L1 compelled her to focus on manner of motion and thus formulate a growth point in which motion is made verbally and imagistically salient, but where the L2 did not offer a readily accessible complex manner verb such as *tambalearse* 'to tumble'. The researchers point out, however, that most L1 speakers of Spanish in all likelihood would not use this verb to narrate the scene. Indeed, as illustrated in (8), one of the L1 Spanish speakers opted to describe the event through a stative construction accompanied by a manner gesture:

(8) la ensalada [echa **un desastre**]
'the salad [is a disaster]'
MANNER = hand shaking
(Negueruela *et al.* 2004: 135)

According to the researchers, the L2 speaker in (7) compensates for the lack of availability of what for her would have been a suitable motion verb[8] by describing the position of the salad, 'in mid-air' synchronized with a manner gesture to depict its motion.

Discussion

While path of motion is marked differently in V-languages compared to S-languages, it is nevertheless marked. As I mentioned earlier, without path there can be no motion. Thus, learning another language is a question of learning how the language encodes path verbally and in gesture. In short, the problem is primarily a linguistic one. Manner of motion, on the other hand, presents a problem of a very different sort – a problem that has to do with acquiring a different conceptual framework for talking and thinking about motion events. Because of the robust inventory in English of fine-grained manner verbs, one would expect English speakers to exhibit enhanced sensitivity to the manner in which motion events occurs, while the same would not hold for Spanish speakers, whose manner repertoire is sparser than English (Slobin 2003). Although Spanish does have ways of encoding manner verbally, the L1 speakers in Negueruela *et al.* (2004) preferred to encode manner, if at all, through gesture only. What all this means for L2 learners is that L1 English speakers would have to 'desensitize' themselves to the manner of motion events and L1 Spanish speakers would have to develop precisely this same sensitivity when learning each other's languaculture. We might even predict that, everything else being equal, L1 Spanish speakers would have an 'easier' time of it when learning English because they would need to take on a new TFS perspective for motion events. English L1 speakers, on the other hand, would have a more difficult task moving to L2 Spanish, since they would have to downplay their need for fine-grained descriptions of manner. This is an intriguing prediction for future research to address.[9]

Conclusion

In this chapter, I have tried to show what the consequences of erasing the Saussurean–Bloomfieldian circle around language are for appreciating the organic connection between language and culture and its relevance for language learning and proficiency. In many ways, languacultures are accumulations of narrowly circumscribed domains that ultimately have a profound impact on how reality is construed. For example, Langacker (2006) shows the consequences of extending the concept of motion to the domain of fictive motion, whereby inanimate objects are metaphorically perceived as being in motion. One has to assume that, if talking and thinking about 'real' motion in a second language is complex, talking and thinking about fictive motion is even more complex and problematic. In English one can say 'The electric cord [goes/runs] from the TV to the outlet' while in Spanish only the first option is possible. In English one can say, 'As I painted the ceiling, paint spots progressed across the floor,' but to render this same meaning in Spanish requires a different way of thinking about the event: *Mientras pintaba el techo, iban cayendo manchas de pintura por todo el suelo e iban formando progresivamente una hilera.* 'As I painted the ceiling, drops of paint were falling on the floor and they were progressively forming a line.'

Another area where meaning comes to the fore with regard to languacultures that I was unable to consider in this chapter is conceptual metaphors. These play a central role in shaping our understanding of reality (see Kövecses 1999 and 2000). Thus, in Western communities we generally conceptualize anger as HEATED LIQUID IN A CONTAINER, which results in linguistic expressions such as 'blowing one's top' and 'letting off steam'. In Chinese, heat does not seem to play much of a role in how anger is conceptualized. Instead, it is metaphorized as EXCESS *QI* (energy) IN THE BODY, giving rise to such linguistic expressions as 'having too much *qi* in one's heart' and 'keeping one's *qi*' (Kövecses 2000: 151). Research to date has shown that L2 learners have a very difficult time assigning appropriate interpretations to metaphors in their new language (see Lantolf and Thorne 2006 for a review of this research), let alone using these concepts to produce meaning. It is therefore imperative to find ways to help learners develop the capacity to interpret and generate meanings that are appropriate in terms of the relevant languaculture; that is, helping them appreciate the significance of, and the learning opportunities provided by, rich points.

Much has been written and said about the ultimate attainment question in L2 learning over the past ten to fifteen years (see Birdsong 2004 for an excellent review of research on this topic). From the perspective of inside-the-circle linguistics, this has meant investigating the extent to which learners achieve native or near-native competence in the grammar and the phonology of an L2 and where little attention is given to the meanings generated by learners. Indeed, as Byrnes (2002: 45) suggests, learner-centred pedagogy has been by and large concerned 'with learners "creatively" expressing personal meanings or applying their own strategies and styles' when using the L2. Once the Saussurean–Bloomfieldian circle is erased, however, and we move into the domain of languaculture, ultimate attainment is about much more than grammatical and phonological ability. It entails 'acquiring the richness of the L2 system's symbolic resources' and a concern with learner ability to make acceptable choices 'within the nexus of intended meaning, available resources, and privileged forms of expression as the L2 speech community has evolved them' (Byrnes 2002: 45).

The new orientation to learning and proficiency necessitates a reconceptualization of the relationship between learners and the language they are learning. Currently, the dominant pedagogical approaches bring language to learners by first rupturing the language–culture nexus that is languaculture and then reducing language, now understood as form and lexical equivalents, to its elemental components, which are doled out to learners in bite-sized chunks, usually at the expense of conceptual knowledge. Thus, in teaching a language such as Spanish, we separate tense (often mistakenly assumed to be synonymous with time) from aspect and then proceed to further dissect aspect into perfect and imperfect, which we present to learners often in separate lessons through a rules-of-thumb account (e.g., use preterite when talking about completed events, such as signalled by temporal adverbs – 'yesterday', 'last year'; use imperfect when describing what was happening or to tell time, to describe a mental or physical state in the past, etc. (see Whitney 1986: 108). Once we reunite them and adopt the perspective of languaculture things become not only more coherent but it becomes imperative to bring the learners to the languaculture. Conceptual understanding becomes paramount not only with regard to metaphors, schema, lexical networks and the like, but also with regard to the conceptual meanings imparted by the grammatical features of a language. Tense then relates to aspect, and perfect and imperfect aspect form a unified way of taking a temporal stance on different events (see Negueruela 2003 and Negueruela and Lantolf 2006). Rich points between different languacultures become the focus of our pedagogical attention as we seek to help students recognize, cope with and use them as the means for developing new ways of understanding reality.

Notes

* I would like to thank Karen Johnson for her helpful comments and feedback on an earlier version of this chapter.

1 It is, perhaps, unfortunate that Whorf himself did not use the term languaculture. If he had, a great deal of confusion and misunderstanding about his claims might well have been avoided. Be that as it may, Agar's term, although cumbersome, appropriately captures the language/culture unity that Whorf insisted on as the proper object of study of linguistics.

2 Lucy and Wertsch (1987) do not use the term languaculture, but I have taken the liberty of using it throughout this discussion to drive home Agar's point that for Whorf language and culture are inseparable.

3 The scholars were born a year apart, Vygotsky in 1896 and Whorf in 1897, and both died at a young age; Vygotsky in 1934 and Whorf in 1941. Although there is evidence that Vygotsky was familiar with the work of Boas, Levy-Bruhl and even Sapir, there is none that he was aware of Whorf's research, nor is there evidence that Whorf was aware of Vygotsky's developing theory (Lucy and Wertsch 1987).

4 The claim that the forms of language developed in modern societies does not imply that they are superior to forms developed in indigenous communities. As Vygotsky and Luria point out, people from modern societies would have a very difficult time orienting themselves to life in indigenous societies and vice versa. For example, while space and distance are exceedingly important in indigenous societies, time and causality tend to be more central in modern societies (Luria and Vygotsky 1992: 65). What is more, Vygotsky seems to have recognized that forms of thinking that occur in indigenous societies also occur with regularity in the everyday thinking of modern technological societies. For a discussion of this topic, see Hallpike (1979).

5 The English lexicon is by no means the richest when it comes to manner verbs.

Luria and Vygotsky (1992: 68) point out that one language (not named) examined in Levy-Bruhl's research had 33 verbs just to describe walking.

6 For a fuller treatment of the studies, see Lantolf and Thorne (2006).

7 While Turkish is a V-language, unlike Spanish it marks manner and path with separate verbs. Other languages that pattern like Turkish are Korean and Farsi. More research is required on gesture/speech synchronization in such languages.

8 According to the researchers, it would have been acceptable in Spanish had the speaker used the verb *moverse*, which is a common Spanish verb that an advanced speaker is likely to know. However, if she had already formulated an English-based growth point that demanded a fine-grained manner verb, this would not have been a reasonable means of expressing what she had perceived in the scene.

9 For cross-linguistic consideration of the role of grammatical temporal relations, particularly bounded and unbounded events and tense and aspect, on ultimate attainment, see von Stutterheim and Carroll 2006; similarly, Carroll and Lambert 2006 address the extent to which language-specific preferences in information structure are driven by grammaticized means.

References

ACTFL (2005) *Executive Summary. Standards for Foreign Language Learning. Preparing Learners for the 21st Century*. http://www.actfl.org/files/public/execsumm.pdf.

Agar, M. (1994) *Language Shock. Understanding the Culture of Conversation*. New York: Quill.

Birdsong, D. (2004) 'Second language acquisition and ultimate attainment', in A. Davies and C. Elder (eds), *The Handbook of Applied Linguistics*. Oxford: Blackwell, pp. 82–105.

Bloomfield, L. (1933) *Language*. New York: Holt.

Byrnes, H. (2002) 'Toward academic-level foreign language abilities: reconsidering foundational assumptions, expanding pedagogical options', in B. L. Leaver and B. Shekhtman (eds), *Developing Professional-Level Language Proficiency*. Cambridge: Cambridge University Press, pp. 34–60.

Byrnes, H., Weger-Guntharp, H. and Sprang, K. A. (eds) (2006) *Educating for Advanced Foreign Language Capacities: Constructs, Curriculum, Instruction, Assessment*. Washington, DC: Georgetown University Press.

Carroll, M. and Lambert, M. (2006) 'Reorganizing principles of information structure in advanced L2s: French and German learners of English', in H. Byrnes, H. Weger-Guntharp, and K. Sprang (eds), pp. 54–73.

CASL (2005) *A Call to Action for National Foreign Language Capabilities*. http://www.nlconference.org/docs/White_Paper.pd.

Clark, A. (1998) 'Magic words: How language augments human computation', in P. Carruthers and J. Boucher (eds), *Language and Thought: Interdisciplinary Themes*. Cambridge: Cambridge University Press, pp. 162–81.

Crowley, T. (1996) *Language in History. Theories and Texts*. London: Routledge.

Fauconnier, G. and Turner, M. (2002) *The Way We Think. Conceptual Blending and the Mind's Hidden Complexities*. New York: Basic Books.

Gullberg, M. (2005) 'Transfer at hand: Gestures and topological relations in learner production'. Paper presented at the AILA World Congress. 25 July 2005. University of Wisconsin, Madison.

Hallpike, C. R. (1979) *The Foundations of Primitive Thought*. Oxford: Clarendon Press.

ILR *Interagency Language Roundtable Language Skill Level Descriptions Speaking.* http://www.govtilr.org/ILRscale2.htm.

Kecskes, I. and Papp, T. (2000) *Foreign Language and Mother Tongue.* Mahwah, NJ: Lawrence Erlbaum.

Kellerman, E. and van Hoof, A-M. (2003) 'Manual accents'. *IRAL*, 41: 251–69.

Kövecses, Z. (1999) 'Metaphor. Does it constitute or reflect cultural models?' in R. W. Gibbs, Jr. and G. J. Steen (eds), *Metaphor and Cognitive Linguistics.* Amsterdam/Philadelphia: John Benjamins, pp. 167–88.

Kövecses, Z. (2000) *Metaphor and Emotion. Language, Culture and Body in Human Feeling.* Cambridge: Cambridge University Press.

Kramsch, C. (2004) 'Language, thought and culture', in A. Davies and C. Elder (eds), *The Handbook of Applied Linguistics.* Oxford: Blackwell, pp. 235–61.

Langacker, R. (2006) 'The conceptual basis of grammatical structure', in H. Byrnes, H. Weger-Guntharp and K. Sprang (eds), pp. 17–39.

Lantolf, J. P. and Thorne, S. (2006) *Sociocultural Theory and the Genesis of Second Language Development.* Oxford: Oxford University Press.

Lucy, J. A. and Wertsch, J. V. (1987) 'Vygotsky and Whorf: a comparative analysis', in M. Hickmann (ed.), *Social and Functional Approaches to Language and Thought.* Orlando, FL: Academic Press, pp. 67–86.

Luria, A. R. (1976) *Cognitive Development.* Cambridge, MA: Harvard University Press.

Luria, A. R. and Vygotsky, L. S. (1992) *Ape, Primitive Man and Child. Essays in the History of Behavior.* Orlando, FL: Paul Deutsch Press.

Mayer, M. (1979) *Frog Goes to Dinner.* New York: DIAL.

McNeill, D. (1992) *Hand and Mind. What Gestures Reveal About Thought.* Chicago: University of Chicago Press.

McNeill, D. (2000) 'Introduction', in D. McNeill (ed.), *Language and Gesture.* Cambridge: Cambridge University Press, pp. 1–10.

McNeill, D. and Duncan, S. (2000) 'Growth points in thinking for speaking', in D. McNeill (ed.), *Language and Gesture.* New York: Cambridge University Press, pp. 141–61.

Negueruela, E. (2003) *A Sociocultural Approach to the Teaching-Learning of Second Languages : Systemic-Theoretical Instruction and L2 Development.* The Pennsylvania State University: Unpublished Ph.D. thesis. University Park, PA.

Negueruela, E. and Lantolf, J. P. (2006) 'A concept-based approach to teaching Spanish grammar', in R. Salaberry and B. Lafford (eds), *Spanish Second Language Acquisition: The State of the Art.* Washington, DC: Georgetown University Press, pp. 158–201.

Negueruela, E., Lantolf, J. P., Jordan, S. R. and Gelabert, J. (2004) 'The "private function" of gesture in second language speaking activity: a study of motion verbs and gesturing in English and Spanish'. *International Journal of Applied Linguistics*, 14, 113–47.

Olson, D. R. (1994) *The World on Paper.* Oxford: Oxford University Press.

Özyürek, A. (2002) 'Speech–gesture relationship across languages and in second language learners: implications for spatial thinking and speaking'. *Proceedings of the Annual Boston University Conference on Language Development*, 26, 500–09.

Saussure, F. de (1959) *Course in General Linguistics.* New York: McGraw Hill.

Shore, B. (1996) *Culture in Mind. Cognition, Culture and the Problem of Meaning.* New York: Oxford University Press.

Slobin, D. I. (1996) 'From "thought and language" to "thinking for speaking"', in S. Gumperz and S. Levinson (eds), *Rethinking Linguistic Relativity.* Cambridge: Cambridge University Press, pp. 70–96.

Slobin, D. I. (2003) 'Language and thought online: cognitive consequences of linguistic relativity', in D. Gentner and S. Goldin-Meadow (eds), *Language in Mind: Advances in the Study of Language and Thought.* Cambridge, MA: MIT Press, pp. 157–91.

Stam, G. (2001) 'Gesture and second language acquisition'. Paper presented at TESOL Convention, St Louis, MO, March.

Stutterheim, C. von and Carroll, M. (2006) 'The impact of grammatical temporal categories on ultimate attainment in L2 learning', in H. Byrnes, H. Weger-Guntharp, and K. Sprang (eds), pp. 40–53.

Talmy, L. (2000) *Toward a Cognitive Semantics. Volume II. Typology and Process in Concept Structuring.* Cambridge, MA: MIT Press.

Whitney, S. (1986) *Spanish/English Contrasts. A Course in Spanish Linguistics.* Washington, DC: Georgetown University Press.

Whorf, B. L. (1956) *Language, Thought and Reality: Selected Writings of Benjamin Lee Whorf*, ed. J. B. Carroll. New York: Wiley.

Part II

Description and Pedagogy

4 Languaging, agency and collaboration in advanced second language proficiency[1]

Merrill Swain

Introduction

The title of this chapter has four parts: languaging; agency; collaboration; and advanced second language proficiency. My intention is to foreground the concept of languaging and its importance to the notion of 'advancedness' in second language proficiency. This I will do by discussing the concept of languaging and concretizing it with examples taken from several data sets I have been working with over the last few years. The examples I will be using for illustrative purposes have been selected from studies in which the participants were asked to engage in various language-related activities *together*. What is seen in the examples is their agency in action. In that sense, the second and third parts of my title serve as the context for the examples I have selected.

Several of Vygotsky's insights into the relationship between language and thought serve as the basis of the arguments presented in this chapter. Vygotsky (1978, 1987) argued that the development and functioning of all higher mental processes (cognition) are mediated, and that language is one of the most important mediating tools of the mind. As such, speaking and writing shape and reshape cognition. This shaping and reshaping of cognition is an aspect of learning, and is made visible as learners talk through with themselves or others the meanings they have, and make sense of them. This means that the capacity for thinking is linked to our capacity for languaging – the two are united in a dialectical relationship.

Languaging

For some time now, I have been searching for a word that puts the focus in second language learning on the importance of *producing* language, but which does not carry with it the conduit metaphor (Reddy 1979) of 'output'. Output is a word that evokes an image of language as a conveyer of a fixed message (what exists as thought). Output does not allow at all for the image of language as an activity – that when a person is producing language, what he or she is engaging in is a cognitive activity; an activity of the mind. Individuals use language to mediate cognition (thinking). In other words, as we consider the notion of 'advancedness' in language use, it is too simplistic to think of language as being only a conveyer of meaning. Rather we need to think of language as also being an agent in the making of meaning (for extended discussion of language as a resource within a Hallidayan framework, see Matthiessen, this volume). It has been important for me to find a word which reflects this meaning of language as a cognitive tool.

In my writing, I began to use the words 'verbalizing' and 'verbalization' (see Swain 2000). However, these words, too, have been subject to misinterpretation – misinterpreted because people often assume that 'verbalizing' refers only to speaking, rather than to both speaking and writing. Over time, the word 'languaging' emerged. For me, it conveyed an action – a dynamic, never-ending process of using language to make meaning.

Perhaps it was fortunate (for me) that I did not know anyone else had used this term. This meant I could develop the word's meaning, giving it sense (Vygotsky 1987) as I used it in interaction with others. This is, in essence, an excellent example of what languaging accomplishes. However, I discovered soon enough, that others, some not so recently, had made use of this word. In fact, today (22 December 2005), I Googled 'languaging' and found over 47,300 entries.

For example, in 1979 Lado wrote a paper entitled 'Thinking and "languaging": A psycholinguistic model of performance and learning'. There Lado states that, 'Since English has no generic term to refer globally to the various uses of language, I will use "languaging" for convenience' (1979: 3). Already, he and I differed in our meaning of languaging. I am using it to refer to *producing* language, and, in particular, to producing language in an attempt to understand – to problem-solve – to make meaning. Lado continues:

> Languaging is full linguistic performance in contrast to partial linguistic performance. In partial linguistic performance the attention of the performer is on some element or part of the language in something less than full communicative use. Such is the case when we concentrate on a word, a grammar point, or a pronunciation problem in language learning. In languaging our attention is not on the language . . . (1979: 3)

This is not at all what I mean by languaging! In fact, it is precisely when language *is* used to mediate problem solutions, whether the problem is about which word to use, or how best to structure a sentence so it means what you want it to mean, or how to explain the results of an experiment, or how to make sense of the action of another, or . . . that languaging occurs. As Becker (1991) states, 'Languaging about language is as everyday as languaging about anything else' (1991: 229).

But I am getting ahead of myself here because this is precisely the point I want to end with – that languaging about language is one of the ways we learn a second language to an advanced level. I will start with an example from a non-language learning context: the context of learning is the circulatory system in the biological sciences. Students who engaged in providing explanations to the self (aloud) after each sentence in a text about the circulatory system developed a more accurate, complete and deeper understanding of how the system works and retained that understanding over a longer time compared to students who were not asked to verbalize their explanations, but only read the text over silently several times (Chi *et al.* 1994).

Why might this be? Languaging serves to mediate cognition. Speaking and writing complete and transform thought (Vygotsky 1987). Vygotsky (1978, 1987), Barnes (1992), Wells (1999) and others have argued that speech and writing can serve as a means of development by reshaping experience. Languaging serves as a vehicle through which thinking is articulated and transformed into an artifactual form. Writing about speaking, Smagorinsky (1998) argued that 'the process of rendering thinking into speech is not simply a matter of memory retrieval, but a process through which thinking reaches a new level of articulation' (1998: 172–3). The same is true for writing. Ideas are crystallized. They become available as an object about which questions can be raised and answers can be explored with others or with the self. In other words, languaging is a process which creates a visible or audible product about which one can language further.

From this point of view, then, it is rather interesting to discover that the term 'languaging' also surfaces in the clinical psycho-therapeutic literature. In 1996, Hall completed a Ph.D. dissertation entitled: 'Languaging: The linguistics of psychotherapy. How language works psycho-therapeutically'. In his dissertation, he talks about languaging and consciousness, and the importance in therapy of 're-languaging', that is, of re-cognizing and restructuring one's knowledge by languaging.

This is an aspect of the concept of 'talking-it-through' (Swain and Lapkin 2002) that we have found in our second language learning data – a point I will return to below. It is the 'coming-to-know-while-speaking' phenomenon. That is, while speaking (or writing), we may reach a new or deeper understanding (Swain 2006).

This is not a new idea. Almost 200 years ago, Kleist (cited in O'Connell 1988) wrote a paper entitled 'On the gradual working-out of one's thoughts in the process of speaking'. Kleist's advice at that time, and relevant to the point I am trying to make, is as follows:

> If you want to understand something and can't figure it out by pondering, I would advise you, my dear ingenious friend, to speak of it to the next acquaintance who happens by. It certainly doesn't have to be a bright fellow: that's hardly what I have in mind. You're not supposed to ask him about the matter. No, quite the contrary, you are first of all to tell him about it yourself (1988: 181).

In other words, through the process of talking-it-through – to another, with another or with the self – we may come to a new understanding, a new insight – we develop and learn. Languaging is not simply a process of 'reading off from within' (O'Connell 1988: 182). It is not a 'brain dump'. As Vygotsky (1987: 219) said, '. . . Thought is not merely expressed in words: it comes into existence through them . . . thought finds its reality and form [in language].'

Languaging, as I am using the term, refers to the process of making meaning and shaping knowledge and experience through language. It is part of what constitutes learning. Languaging about language is one of the ways we learn language. This means that the languaging (the dialogue or private speech) about language that learners engage in takes on new significance. In it, we can observe learners operating on linguistic data and coming to an understanding of previously less well understood material. In languaging, we see learning taking place.

Languaging mediates second language learning

In our recent research (e.g., Swain 2005; Swain and Lapkin 1998, 2002, 2006), we have been exploring how languaging is a source of second language learning. Our participants have been young adolescents (11, 12 and 13 years old) who are in French immersion programs in Toronto. Our research paradigm has been to ask students to write a story for which the stimulus is either an audio-recording or a set of drawings. We then reformulate the stories the students have written. In reformulating the stories, the intent has been to *not* change the meaning of what the students wrote, but to change the form of their writing so that it would be acceptable to a fluent user of the target language. We then ask the students, some of whom have written their stories on their own and others with a partner, to notice the differences between their story and the rewritten one. While they are noticing the differences, they are videotaped. We then play back the videotape to the participants, stopping the tape each time the students noticed a difference between their writing and the reformulator's (whom they knew to be a French-speaking adult). We ask the students to tell us what they were thinking, sometimes prompting them with specific questions. During all three stages of their writing (during the original creation, during the noticing and during the stimulated recall), the students engaged in languaging. Initially they mediated their writing by languaging, by working through how best to write their intended meaning. During the noticing stage and stimulated recall stages, the students expressed their beliefs about the target language, often languaging themselves through to an understanding of why the reformulator had changed what they had written. The impact of their languaging shows up when, on their own, they later rewrite what they had originally written, incorporating the substance of what they had languaged about. More proficient students do this more effectively than less proficient ones do (Qi and Lapkin 2001).

Languaging as talking-it-through: cognitive and memorable engagement

Watanabe (2004) conducted a study using a similar research design as Swain and Lapkin (see above), but with Japanese adult learners of English, studying English to be able to enter university. Watanabe was particularly interested in collaborative learning where the participating partners had different levels of English proficiency. In the example I provide here, Ken is the high proficiency student and Yoji is mid-proficiency. It is one of many examples from her data set of students coming to an understanding of a particular, contextualized use of English through languaging.

What will be seen in this example may on the surface appear rather trivial – working out the correct use of the definite article. However, it reflects a difficult and important problem for L2 learners of English. In this case, it is compounded by the prepositional system which is just as troublesome for L2 learners of English. But, as we will see, the data reveal that the real problem is actually working out the constituents in a noun phrase. From a pedagogical perspective, the example shows that diagnosing the problem as only one of article use, or of prepositional use, is rather off the mark: the more general issue is about the constituents in a noun phrase.

As we will see, Ken completely rejects the feedback he receives from the authoritative target language speaker. He only comes to terms with it once he has worked out a solution by talking-it-through. Although he is supported by Yoji, the reality in this example is that Ken, by talking and thereby constructing his thoughts, is able to locate the contradictions and reconstruct his understanding of what is going on. In other words, by languaging, Ken develops a rule that makes what was initially seen as contradictory consistent.

Their task was to read a prompt, taken from a TOEFL writing prompt, and jointly write a response. The prompt read as follows: 'Choose one of the following transportation vehicles and explain why you think it has changed people's lives: automobiles, bicycles, airplanes. Use specific reasons and examples to support your answer.'

Yoji and Ken chose to write about aeroplanes (see Appendix A for the full text of what they wrote). At the beginning of the second paragraph, they wrote, 'First, if I were a person who lived *in nineteenth century of Japan* and wanted to study in America, I could not afford to buy a ticket . . .' At the beginning of the third paragraph, they wrote, 'Secondly people *in nineteenth century* had a big problem about sending their letters to another country.' The reformulator changed 'a person who lived *in nineteenth century of Japan*' to 'a person who lived *in nineteenth-century Japan*', and 'people *in nineteenth century*' to 'people *in the nineteenth century*'. Small points, indeed, especially as Yoji and Ken's intent was clear. But advanced second language learners are expected to know these things.

In the noticing stage, when Yoji and Ken were comparing their own writing with that of the reformulator, Ken read out loud what the reformulator had written, indicating by a pause, followed by 'Ohh' that he did not much like what the reformulator had done. In the following turn, Yoji indicated that he thought it should be '*of* Japan'. As we see in turns 139–154, Ken refuses to accept the reformulator's correction because he already had a well-used pattern of speech that he did not want to give up.

139 K: We don't need anything after century? . . . No, I don't think so! {=I don't think that's right; I think we do need something after century}.
140 Y: . . . who lived in nineteenth century . . . nineteenth-century Japan, in nineteenth century Japan.
141 K: We need 'in' or 'of'. I think so, there {pointing}, 'in' or 'of'.
142 Y: Mmm. {non-committal}
143 K: Because nineteenth century is noun, not adjective. What, what do you think about it?
144 Y: But he (the reformulator) omitted 'of' {willing to accept reformulator's feedback}.
145 K: Mm hmm, mm hmm. {pondering}
146 Y: So . . .
147 K: I don't know why.
148 Y: He wanted to say nineteenth century. {Yoji has accepted the reformulation}
149 K: Like adjective. {=but that's what you'd do if it was an adjective}
150 Y: It is adjective and nineteenth-century Japan? {Yoji tries, timidly, to say that nineteenth century *is* an adjective, an adjective that modifies Japan}
151 K: I think this one is . . . this sentence . . . nineteenth century of Japan . . . is correct, not, not wrong. I think so. Or maybe 'in'?
152 Y: But we . . .
153 K: Because I always use this kind of phrase.
154 Y: Mm hmm {Yoji decides it's time to move on}.

This is an excellent example of Ken exercising his agency. It is also an excellent example of what happens to feedback: why the corrective feedback, the input provided by the reformulator did not 'work' for Ken. Ken insists on treating 'nineteenth century' as a noun phrase rather than the adjective phrase it is. Ken's response to the feedback was mediated by his past learning ('I always use this kind of phrase' (turn 153)), and reinforced by his present languaging ('Because nineteenth century is noun, not adjective' (turn 143)). Note that the assumption made by many theoreticians, researchers and teachers is that, if the feedback is noticed, depending on various feedback characteristics (e.g., salience, length, complexity, linguistic feature involved), it will be 'taken in' to the learner's developing competence. This assumption places the responsibility on the feedback, not with the learner.

A sociocultural perspective is quite different. It assumes the environment provides the opportunities for learning, but that it is the learner, with his or her history, in his or her immediate environment, who has options and makes choices. This is the learner as *agent*: as an individual who perceives, analyses, rejects or accepts solutions offered, makes decisions and so on. This is the learner as an 'agent-operating-with-mediational means' (Wertsch 1998: 26).

Many turns following the excerpt above, Yoji and Ken noticed that the reformulator had added 'the' to 'in nineteenth century'. Through languaging, Ken reasons through to an understanding that explains both the changes made by the reformulator: 'nineteenth century' is a noun phrase in the second usage and so the article 'the' is needed. Because it is not needed in the first use of 'nineteenth century', then it must be that the first use of 'nineteenth century' was as an adjective. And, as an adjective, it modifies 'Japan' directly and 'of' is therefore not needed.

264 Y: 'People in the' . . . {reading}
265 K: in the, in the, in the, in THE nineteenth century.
266 Y: Here, 'in nineteenth-century Japan'. {referring to the first reformulation}
267 K: Ahhh! {the moment of insight}
268 Y: So this is a different . . . so if we put 'the'.
269 K: Yeah, it sh-, it should be noun, noun. {= we should put 'the' if 'nineteenth century' is a noun}
270 Y: In the nineteenth century.
271 K: If we, if we, if we want to use ' nineteenth century' as a noun . . .
272 Y: Mm.
273 K: . . . maybe we need article.
274 Y: Article. If we don't put articles . . .
275 K: We don't have to put in article for 'in nineteenth-century Japan' because this 'nineteenth century' is adjective . . . difference. Okay.

In the post-test they rewrote on their own their original story. Ken adhered to his new-found, languaged insight, rewriting 'in nineteenth century of Japan' as 'in nineteenth-century Japan', and 'in nineteen century' as 'in the nineteenth century'. Yoji, although the one to encourage Ken on, had himself not benefited from Ken's explanation and did not distinguish between the two uses of 'nineteenth century': he omitted the article 'the' in both cases. What Ken did here was to reverse his adamant rejection of the reformulator's feedback by languaging: through it, he was able to focus on an apparent inconsistency in language usage, reason about it and reconcile it. He made use of prior knowledge, but he also created new knowledge – for himself – in the process.

Languaging as talking-it-through: affective and memorable engagement

An aspect of second language learning that is a major challenge to advanced level second language learners is the use and understanding of humour. The use and understanding of humour requires high levels of not only linguistic knowledge, but social and cultural knowledge as well. According to Tocalli-Beller (2005), 'L2 humor has earned the reputation of [being] "unteachable" (and even unnecessary and frivolous), prompting L2 teachers to shun its inclusion in the curriculum' (2005: 37–8). Yet, humour is very much part of everyday life (Carter and McCarthy 2004; Nerlich and Clarke 2001). As Cook (2000) states:

> Knowing a language, and being able to function in communities which use the language, entails being able to understand and produce play with it, making this ability a necessary part of advanced proficiency. (2000: 150)

Tocalli-Beller's Ph.D. dissertation research (2005) is one of the first studies to be carried out on the *intentional* inclusion of humour in an L2 curriculum. Tocalli-Beller designed and taught a course for international students, most of whom were studying at the graduate level. The material she used in her course consisted of a selection of jokes, cartoons and riddles from national newspapers. The design of her study allowed her to trace in the languaging (peer–peer dialogues) *about* language how these ESL students moved from non-comprehension to spontaneous use.

During the eight-week course students interacted together, first in pairs to understand the jokes, riddles and cartoons that they had been given; and then, secondly, with the rest of the class to tell their jokes to them. The students were also given a series of pre- and post-tests related to the 'semantic triggers' – what needed to be understood for the joke to be a joke. For example, the policeman says, 'You can't park here.' The driver asks, 'Why not?' The policeman says, 'Read that sign.' The driver says, 'I did. It says, "Fine for parking." ' In order to understand the humour, you need to know the two meanings of 'fine'.

Or, to answer the riddle 'What do you get if you cross a snowman and a vampire?' with 'Frost-bite', one needs to know the word 'frost-bite'.

The students were tested on their knowledge of the semantic triggers, using Wesche and Paribakht's (1996) Vocabulary Knowledge Scale (VKS). 'The scale ratings range from complete unfamiliarity, through recognition of the word or some idea of its meaning, to the ability to use the word with grammatical and semantic accuracy' (1996: 29). Tocalli-Beller adapted the VKS so that it tapped knowledge of different meanings of the same word. The students were pre-tested at the start of the course on their knowledge of *all* the semantic triggers used in the course materials; then each pair was post-tested soon after their activities in pairs on only the semantic triggers that appeared in their jokes. Finally, after the (first round of) whole-class activities, and again at the end of the course, the students were post-tested on their knowledge of *all* the semantic triggers (see Table 4.1).

Harry and Will, one of the pairs of students in Tocalli-Beller's class, were given a set of jokes, which included a riddle. The riddle was: 'Why did the dentist and manicurist divorce?' The answer – and the semantic trigger – is 'Because they fought tooth and nail'. Much of the languaging these students

Table 4.1 Languaging: A microgenetic analysis for 'to fight tooth and nail' (Tocalli-Beller 2005)

ST	*To fight tooth and nail*	
Pair	Harry	Will
Pre-Test Week 1	I don't remember having seen this expression before.	I don't remember having seen this expression before.
Riddle	Why did the dentist and the manicurist divorce? Because they fought tooth and nail.	
Riddle Activity Start of Week 2	289. H: Tooth and nails means uh . . . Oh, I mean, tooth and nail, I mean, when two persons or uh, like they quarrel too bad, I mean. They have like the same, like the same expression of my language. 290. W: For tooth and nail? 291. H: Tooth and nail? Well, not the same [while flicking the pages of the dictionary]. Not the same tooth and nail but uh, I don't know. 292. W: Tooth. 293. H: You see, it's uh it's treat bad. 294. W: Mmmm. . . . 295. H: The persons, they fought uh with uh— 296. W: [laughs] they fight with tooth and nails. They fight with 'a lot of effort or determination to do something. We fought tooth and nail to get our plans.' 297. H: Yes, when you want something very much. You try very hard to get it.	
Post-Test End of Week 2	To try hard with all power to get something. He has fought tooth and nail to get this job.	Try hard. Work hard.
Stimulated Recall Week 3	192. W: Yes, fight tooth and nail. 193. R: And what does that mean? 194. W: Very hard. (. . .) 195. H: Because in Persian we have, I wanted to translate in English . . . we have the same thing claws and tooth. With claws and tooth. He tries very hard with all his means.	
Class Activity Week 2	228. Will: Dentist takes care of what? 229. Eric: *The tooth.* 230/1. Lisa and John: *The teeth.* 232. Will: And the manicurist take care of what? (. . .) 235. Helen: *Fingers.* 236. Harry: Not fingers. 237. Don: Nails. Nails. (. . .) 244. Harry: Yes, you must know an expression in English. 245. Lisa: *Fight tooth and nail.*	

Table 4.1 *continued*

	246. John: *To fight with tooth and nails!* (. . .) 250. Lisa: [to John] I think we got it [laughs] (. . .) 256. John: Yes, there is a phrase that you fight very hard. So you just give anything you can give, right? to fight for what they need. (. . .) 276. Harry: I fought tooth and nail to get a promotion. 277. Eric: – Tooth or teeth. 278. Don: *Fought* 279. Eric: *Teeth, I think.* 280. Don: Teeth? 281. Lisa: No, tooth. Tooth and nail. 282. Eric: Tooth and nail. 283. Don: Fought. Fought? 284. Harry: Fight [pauses] Fought. 285. Don: Fight. Fought. OK.	
Class Activity Post-Test Week 3	To try hard and by all means. He fights tooth and nail to pay his depts [debts].	Work very hard to get it.
Delayed Post-Test Week 10	To try hard and by all means to catch something.	Work and try hard.

ST= semantic trigger: key word/expression or centre of energy where the whole matter of the joke is fused.
Italic – spoken simultaneously
(. . .) some turns omitted

did to solve the riddle can be found in Table 4.1 (some turns are omitted to save space).

Neither Harry nor Will knew the meaning of the semantic trigger, in this case an idiomatic expression, 'to fight tooth and nail'. In the second week of the course, when they were given the riddle, Harry struggled to give the meaning of an expression that he thinks is similar in his own language (Persian). In turn 289, we see him saying: 'Tooth and nails means uh . . . Oh, I mean, tooth and nail, I mean, when two persons or uh, like they quarrel too bad, I mean. They have like the same, like the same expression of my language.' And in response to Will's question, Harry (in turn 291) states, 'Well, not the same. Not the same tooth and nail but uh, I don't know.' While Harry is struggling with the meaning of 'tooth and nail', he is looking up the expression in the dictionary, finds it and points it out to Will, saying, 'You see, it's uh it's treat bad' (turn 293). In turn 296, Will 'gets it' as he reads out loud the dictionary definition and the example given: 'They fight with "a lot of effort or determination to do something. We fought tooth and nail to get our plans."' and in turn 297, Harry restates the definition in his own words: 'Yes, when you want something very much. You try very hard to get it.'

In the first post-test, Harry and Will are both able to give a definition of 'tooth and nail'. Perhaps because his languaging has retrieved a similar expression in his first language, Harry is able to create a sentence using it. This possibility is supported by what Harry says during the stimulated recall session when the tape of their original dialogue is played back to them. Harry explains: 'Because in Persian we have, I wanted to translate in English . . . we have the same thing, claws and tooth. With claws and tooth. He tries very hard with all his means' (turn 195).

In the class activity, after asking the riddle question, Will gives a series of hints such that Lisa and John get the answer (turns 228–50). In turn 276, Harry demonstrates his understanding of this newly learned idiomatic phrase by using it in a sentence: 'I fought tooth and nail to get a promotion.' This leads to further languaging as students work out whether it should be tooth or teeth, and fight or fought (turns 277–85).

As further evidence of the power of the students' languaging, Don, another student in the class, after starting to present the riddle that his group had talked through, says, laughingly, 'I fought tooth and nail to get this joke', and the entire class laughed. Much later, in a final interview, Don commented on the memorability of his learning:

> It's much more easy to remember because it has emotional component. Then I sort of remember the situation. I remember who said what. This as a discussion group is great. It's great. Because all the words I could find in my mind after this lesson, you know. Now I remember. So for me it's very useful.

In the post-test held after the whole class activity had taken place, Harry and Will continued to define 'tooth and nail' correctly, and Harry's sentence showed that he was able to transfer his languaged meaning to a new, although probably related, context – from 'getting a promotion' during the class activity to 'paying his debts' in the post-test. In the final post-test, eight weeks after the first time they discussed the expression, both Harry and Will continued to provide the meaning they had learned.

Conclusion and discussion

In the examples I have described, the students were stimulated to language about language because in doing the tasks, they realized that there were things about the target language that they did not know, or were unsure of. Faced with this, they set about trying to solve the problem, using language as a tool to mediate their thinking (cognition). In the studies I have cited, learning has taken place between pre- and post-tests. In the examples I have provided, the students' post-test results are directly traceable in, and to, the dialogue – the languaging – of the students. Languaging mediated the students' language learning by drawing their attention to language-related problems they had, and by giving them the tools to reason with, to solutions.

What Harry and Will and Yoji and Ken's languaging has accomplished for them is two fold. First, their languaging articulated and transformed their thinking into an artifactual form, and as such it became available as a source of further reflection. Secondly, languaging was the means of that further reflection. Through it, these students created new meanings and understandings – that is, they learned both through and about language.

With advanced language learners, our conceptualization of language learning and L2 language use must address the relationship between language and thought. I have argued in this chapter, based on the writings of Vygotsky, that thinking is intimately related to language. Vygotsky argues that higher mental processes find their source in interaction between an individual, others and the artifacts they create, and that the process of interaction is mediated by psychological tools, of which language is one of the most important. Speaking and writing, Vygotsky argued, do much more than convey a message. They serve as tools of the mind, mediating the cognition and re-cognition of experience and knowledge.

Note

1 This chapter is a revised version of a plenary address that was given at the Georgetown Round Table Conference in March 2005. I would like to thank the many people who languaged with me even before this chapter was conceived, and who have read and commented on earlier versions of it: Heidi Byrnes, Ping Deters, Huamei Han, Li-Shih Huang, Yas Imai, David Ishii, Penny Kinnear, Jim Lantolf, Sharon Lapkin, Linda Steinman, Wataru Suzuki, Harry Swain, Steve Thorne, Agustina Tocalli-Beller and Yuko Watanabe.

References

Barnes, D. (1992) *From Communication to Curriculum* (2nd edn). Portsmouth, NH: Heinemann.

Becker, A. L. (1991) 'A short essay on languaging', in F. Steier (ed.), *Reflexivity: Knowing as Systemic Social Construction.* Newbury Park, CA: Sage, pp. 226–34.

Carter, R. and McCarthy, M. (2004) 'Talking, creating: interactional language, creativity, and context'. *Applied Linguistics*, 25, 62–88.

Chi, M. T. H., de Leeuw, N., Chiu, M-H. and LaVancher, C. (1994) 'Eliciting self-explanations improves understanding'. *Cognitive Science*, 18, 439–77.

Cook, G. (2000) *Language Play, Language Learning.* Oxford: Oxford University Press.

Hall, L. M. (1996) 'Languaging: The linguistics of psychotherapy. How language works psycho-therapeutically: An exploration into the art and science of "Therapeutic languaging" in four psychotherapies using general semantic formulations'. Unpublished Ph.D. Dissertation, The Union Institute. Dissertation Abstracts International A, 57(11), 4717.

Lado, R. (1979) 'Thinking and "languaging": a psycholinguistic model of performance and learning'. *Sophia Linguistica*, 12, 3–24.

Nerlich, B. and Clarke, D. (2001) 'Ambiguities we live by: towards a pragmatics of polysemy'. *Journal of Pragmatics*, 33, 1–20.

O'Connell, D. C. (1988) *Critical Essays on Language Use and Psychology*. New York: Springer-Verlag.

Qi, D. S. and Lapkin, S. (2001) 'Exploring the role of noticing in a three-stage second language writing task'. *Journal of Second Language Writing*, 10, 277–303.

Reddy, M. (1979) 'The conduit metaphor', in A. Ortony (ed.), *Metaphor and Thought*. Cambridge: Cambridge University Press, pp. 164–201.

Smagorinsky, P. (1998) 'Thinking and speech and protocol analysis'. *Mind, Culture, and Activity*, 5, 157–77.

Swain, M. (2000) 'The output hypothesis and beyond: mediating acquisition through collaborative dialogue', in J. Lantolf (ed.), *Sociocultural Theory and Second Language Learning*. Oxford: Oxford University Press, pp. 97–114.

Swain, M. (2005) 'The output hypothesis: theory and research', in E. Hinkel (ed.), *Handbook of Research in Second Language Teaching and Learning*. Mahwah, NJ: Lawrence Erlbaum, pp. 471–84.

Swain, M. (2006) 'Verbal protocols: what does it mean for research to use speaking as a data collection tool?', in M. Chalhoub-Deville, C. Chapelle and P. Duff (eds) *Inference and Generalizability in Applied Linguistics: Multiple Research Perspectives*. Amsterdam/Philadelphia: John Benjamins pp. 97–113.

Swain, M. and Lapkin, S. (1998) 'Interaction and second language learning: two adolescent French immersion students working together'. *Modern Language Journal*, 82, 320–37.

Swain, M. and Lapkin, S. (2002) 'Talking it through: two French immersion learners' response to reformulation'. *International Journal of Educational Research* (Special issue on the role of interaction in instructed language learning), 37, 285–304.

Swain, M. and Lapkin, S. (2006) '"Oh, I get it now!" From production to comprehension in second language learning', in D. M. Brinton and O. Kagan (eds), *Heritage Language Acquisition: A New Field Emerging*. Mahwah, NJ: Lawrence Erlbaum.

Tocalli-Beller, A. (2005) 'Peer–peer Dialogue: Bringing Second Language Learning into Play'. Unpublished Ph.D. dissertation, OISE, University of Toronto.

Vygotsky, L. S. (1978) *Mind in Society: The Development of Higher Psychological Processes*. Cambridge, MA: Harvard University Press.

Vygotsky, L. S. (1987) *The Collected Works of L. S. Vygotsky. Volume 1. Thinking and Speaking*. New York, NY: Plenum Press.

Watanabe, Y. (2004) 'Collaborative Dialogue between ESL Learners of Different Proficiency Levels: Linguistic and Affective Outcomes'. Unpublished M.A. thesis, OISE, University of Toronto.

Wells, G. (1999) *Dialogic Inquiry: Toward a Sociocultural Practice and Theory of Education*. Cambridge: Cambridge University Press.

Wertsch, J. V. (1998) *Mind as Action*. New York: Oxford University Press.

Wesche, M. and Paribakht, S. (1996) 'Assessing second language vocabulary knowledge: depth versus breadth'. *Canadian Modern Language Review*, 53, 13–40.

Appendix A

Yoji and Ken from Watanabe (2004) – their written text

Before industrial revolution began, people had to use a horse wagon or their own foot for transportation. Such new transportation as automobiles, bicycles and airplanes improved people's transportation. It is certain that all of them are important but what made people's life more convenient is airplane. Although people could go abroad by ships before airplane was invented, it took too long time and expensive for ordinal people to use. Airplane improved these problem. We would like to give some examples to support this idea.

First, if I were a person who lived *in nineteenth century of Japan* and wanted to study in America, I could not afford to buy a ticket of ship because of price. So, I need Japanese government's help and it might take about one month. It was a big trip in our life; however, now we can go to America cheaper and faster by airplane. That is why many Japanese can go to America and study there. Studying abroad becomes very common and popular for the benefit of airplane.

Secondly people *in nineteenth century* had a big problem about sending their letters to another country. If my parents wanted to let me know about their problem such as relative's death, I would know about their unhappiness after one month. In contrast to that tragedy, now I can know about their problem in a few days by express airmail service. We might be able to go their funeral. Airplane improved system of mailing.

To summarise, important things in our life such as studying abroad, mailing system, and transportation become very convenient because of airplane. Nowadays we can go everywhere in the world in a day, which could not be imagined before airplane was invented.

5 Grammar as a resource for the construction of language logic for advanced language learning in Japanese

Kazuhiro Teruya

Thinking 'grammatically': learning through language to advance language learning

This chapter investigates the role of grammar for instructed intermediate to advanced adult learners and exemplifies that role in the context of learning Japanese as a second/foreign language (henceforth SL/FL). Following Halliday (1978) I see language as a meaning-making resource and consider the fact that language has the potential for creating linguistic meaning as arising from the presence of a grammar within the content systems of language (see Matthiessen, this volume). Grammar makes language an open dynamic system capable of creating rather than merely reflecting meaning. Grammar does this, in the first instance, by theorizing about human experience and enacting human relationships and also by enabling the integration of human knowledge and interaction in the form of spoken or written discourse (Halliday and Matthiessen 1999).

Learning is a process of constructing a network of relations of meanings modelled as commonsense knowledge in the language of everyday life (such as 'survival' Japanese) and as expert knowledge in the language of the disciplines (such as scientific knowledge, see Halliday and Martin 1993). All knowledge is about knowing something; and 'to "know" something is to have transformed it into meaning, and what we call "understanding" is the process of that transformation' (Halliday 2002: 8). In SL/FL learning, the knowledge one has already construed in a language one has mastered is processed and transformed into, or 'mapped onto', an entirely new mode of meaning in the target language, a shift in mode that affects all aspects of the language to be learnt. Following the principles of systemic functional theory (SFL) (for an overview, see Matthiessen, this volume), this transformation into 'meaning' is a semiotic process, a 'semogenesis', which is powered by the semogenic energy of language that inheres in grammar (Halliday and Matthiessen 2004). It is grammar that provides the power to create the networks of semantic relations that construe knowledge in conscious learning.

Specifically, I will explore how learners might come to know and understand (rather than only speak and listen to) Japanese in SL/FL advanced learning contexts. My focus is on the semogenesis of knowing and understanding Japanese with particular reference to the development of the natural logic of the language, the logic that has evolved over countless generations of speakers as part of the evolution of language and that stands in contrast with the designed systems of modern mathematical and symbolic logic (cf. Halliday and Matthiessen 1999). As already indicated, that semogenesis of knowing and understanding the target language is made possible through learning to 'think grammatically': Learning a SL/FL is then the process by which learners become empowered to use the grammar of the L2 consciously as a tool for thinking with, and therefore, for knowing and understanding how the language works. To illustrate how thinking grammatically takes place in the context of Japanese language teaching and learning, I will refer to evidence from journals in which students explore their own processes of learning the grammar by writing about them as part of the requirements of two consecutive optional courses offered to advanced Japanese language learners and native speakers of Japanese; the course, entitled 'Discover Japanese Grammar', is taught at the University of New South Wales in Sydney, Australia.

The domain of linguistic inquiry is that of clause complexing, the combining of clauses by means of logico-semantic relations such as restatement, addition, time, condition and cause in the construction of rhetorical organization in discourse (for overviews of work on this area, see Bybee and Noonan 2002, Haiman and Thompson 1988; for systemic functional interpretations, see Halliday and Matthiessen 2004: Chapter 7, Matthiessen 2002, Teruya, in press: Chapter 6; for a systemic-functional account of the relationship between clause complexing in grammar and rhetorical organization in semantics, see Matthiessen and Thompson 1988). The following example of a clause complex taken from a passage of narrative text that was used and analysed in the course provides a first indication of the issues in focus. The excerpt is extracted from a children's story called *Hanasakaji* 'Grandpa, the bloom bloomer'. It represents a scene where a dog called Shiro is forced to go to the mountain by grandparents from next door. The clause complex consists of nine clauses and each clause is analysed logically in order to show its interdependency relations. The table is organized as follows: column 1 indicates the clause number; columns 2–4 provide a full logical analysis of the clauses; column 2 represents the linear sequences of the clause nexus that are nested in layers; column 3 indicates the value of dependency of each clause that is inherited from the previous nexus, where Greek letters express dependent clauses and Roman numerals independent clauses and added notations indicate logico-semantic relations: elaborating =, enhancing x, and locution "; column 4 indicates subtypes of logico-semantic relations; finally, column 5 provides clause examples.

The example illustrates the contribution clause complexing makes to the development of the narrative episode quoted above. More generally, it illustrates how clause complexing represents an important gateway between the

Table 5.1 Clause complex consisting of more than two clauses

#	logical analysis: logical				dependency inheritance	systemic option selected	examples
1	$\beta^{=}$				β	'paradoxical'	*Tonari no ojiisan to obaasan wa, Shiro ga nani mo iwanai noni* Grandpa and Grandma from next door, although Shiro hasn't said anything
2	α	$\beta^{\times}$	$\delta^{\times}$		$\alpha\beta^{\times}\delta^{\times}$	temporal: progressive	*kura o tsukete nee* put a saddle on him, and
3			$\gamma^{\times}$		$\alpha\beta^{\times}\gamma^{\times}$	temporal: progressive	*kamasu o tsukete nee* attached a straw bag, and
4			$\beta^{\times}$		$\alpha\beta^{\times}\beta^{\times}$	temporal: progressive	*kuwa o tsukete nee* attached a hoe, and
5			α		$\alpha\beta^{\times}\alpha$	temporal: progressive	*sono mata ue ni, ojiisan to obaasan ga notte* and on top of that, Grandpa and Grandma both climbed on:
6		α	1"	1=	$\alpha\alpha$1" 1=	locution: imperative	*"Sore ike,* "Go ahead.
7				2	$\alpha\alpha$1"2	locution: imperative	*yama e ike" to,* Go to the mountain,"
8			2	$\beta^{=}$	$\alpha\alpha2\beta$=	exemplifying: exhaustively	*donattari* they shouted and
9				α	$\alpha\alpha2\alpha$	exemplifying: exhaustively	*tataitari shimashita* whacked (Shiro).

'Without Shiro asking, Grandpa and Grandma from next door put a saddle on him, attached a straw bag, attached a hoe, and on top of that, Grandpa and Grandma from next door both climbed on: "Go ahead. Go to the mountain," they shouted as they whacked Shiro.'

semantics of text and the grammar of the clause, in itself a good reason for focusing on this domain in the context of advanced SL/FL education. Another justification is that clause complexing has generally not been taught systematically even though it can be expected to help expand learners' resources for reasoning in the SL/FL and for developing rhetorical patterns in discourse. This omission is all the more surprising as the ability to generate and understand language logic clearly distinguishes advanced learners from less capable language users. One way to explain that lacuna is that the teaching of clause complexes is approached either from a structural perspective, for instance with reference to structural conjunctions, or from a discourse perspective, for instance with reference to meaning 'implied' in the discourse. On a deeper level, that pedagogical and learning failure also comes about because no appropriate account of clause complexing has been developed and presented to learners, which leaves them with a significant gap between low-level structural patterns and high-level rhetorical patterns. But learners need an account that deals with the overall system of 'natural logic' and the underlying mechanism (or algorithm) of the workings of clause complexing, an account that would show not only how patterns of natural logic are constructed dynamically as language users produce and interpret text, but also how other metafunctional strands of meaning are unified or 'woven together' as clauses are combined into clause complexes.

In the following discussion, I will first illustrate how the meanings of clause complexes are identified in SFL and dynamically constructed in the grammar in relation to the other metafunctions, in particular, the interpersonal system of MOOD, the experiential system of PROCESS TYPE and the textual systems of CONJUNCTION and THEME. In the next step I will show how learners' grammatical knowledge of the workings of clause complexes opens up the world of discourse for them as they move from clause complexes through the rhetorical organization of text.

The approach to explicating the construction of natural logic in language is based on three principles: (1) it is theory-based, inasmuch as it uses the global and local dimensions of a systemic functional theory of language (for an overview, see Matthiessen, this volume); (2) it is description-based, in that the account draws on a comprehensive systemic functional account of the grammar of Japanese (Teruya 2004, in press); and (3) it is learner-centred, in that theory and description not only guide learners to advance their language learning but enable their self-learning of the grammar: their learning journals reveal their reasoning about the process of creating meanings, that is, their being engaged in 'grammatical thinking'.

Language logic for intermediate to advanced Japanese language learning

In SL/FL teaching and learning logical relations construed by the grammar of clause complexing are often dealt with at a very low level of complexity, such as the nexus (combination) of two clauses being linked by some logico-semantic relation, e.g., cause: *ame ga hutte iru node,* || *dekakemasen*[1] 'because it's raining, I won't go out'. However, in naturally occurring texts of the kind advanced learners have to engage with, clause complexes often extend far beyond bi-clausal nexuses such as those presented in Table 5.1 (cf. Matthiessen 2002: 254 for relative frequencies of recursion leading to multi-clause complexes in English).

Consequently, the common approach to teaching clause combining in intermediate to advanced Japanese language classes fails to deal with the 'emergent' grammatical property of clause complexes (Matthiessen 2002: 245), namely, the system of 'recursion', which allows a clause complex to expand repeatedly beyond the minimal bi-clausal nexus. Indeed, this possibility of 'looping back' is one of the inherent grammatical properties of clause complexes. Hence, Matthiessen notes, 'clause complexes are open-ended rather than pre-defined structures' (1995: 140). Furthermore, this dynamic, open-ended nature of clause complexes offers an optimal linguistic environment for demonstrating how linguistic systems are 'unified' dynamically in the construction of language logic. Adopting an approach that deals with this logical system of recursion is an important departure from much SL/FL learning, inasmuch as it replaces a pre-defined, static clause complex structure with the rhetorically oriented language logic that is created dynamically in discourse. In SL/FL learning, therefore, learners are no longer left on their own to deal intuitively with coding the very clause complexes that are engendered as discourse unfolds; instead, they are empowered to deploy grammatical resources in order both to analyse passages of text and to think critically about the role of language in discourse. Halliday describes that shift in the following fashion:

> The value of having some explicit knowledge of the grammar of written language is that you can use this knowledge, not only to analyse the texts, but as a critical resource for asking questions about them: why is the grammar organized as it is? why has written language evolved in this way? what is its place in the construction of knowledge, the maintenance of bureaucratic and technocratic power structures, the design and practice of education? You can explore disjunctions and exploit for potential for creating new combinations of meanings. (1996: 350)

In what follows, I will address the value of having conscious knowledge of the recursion of clause complexes in SL/FL learning with respect to other grammatical systems, particularly with respect to the interpersonal and textual functions of Subject and Theme and the role of conjunctions in the logical systems of TYPE OF INTERDEPENDENCY and TAXIS. I will also refer to the experiential system of PROCESS TYPE. The illustration includes various instances of grammatical thinking that have taken place in the mind of learners as forms of FL/SL learning.

Language logic: metafunctional contribution at the level of clause complexes

The system of CLAUSE COMPLEXING provides 'the strategies open to speakers in developing clause complexes and to listeners in tracking and reconstructing clause complexes' (Matthiessen 2002: 249). Clause complexes represent a sequence of figures (configurations of processes, participants involved in them and attendant circumstances) in a series of interrelated events that are linked logically according to the logico-semantic relationship in which one clause stands to another. Each pair of clauses combined by a logico-semantic relation is called a clause nexus (Halliday 1994: 218); a clause nexus can be further expanded serially by opening up another nexus until such time as the whole complex comes to a full stop (while the text may continue to unfold). The system by which a nexus is dynamically expanded is called the system of RECURSION (with the options of 'stop' versus 'go on').

The nature of the logico-semantic relation by which a clause nexus is formed is determined by the interrelated systems of TAXIS and LOGICO-SEMANTIC TYPE. Respectively, they define (1) the 'degree' of interdependency between the clauses in a clause nexus – either parataxis ('equal status') or hypotaxis ('unequal status'), and (2) the semantic principles according to which clauses are organized into a nexus – either projection (quoting or reporting speech or thought) or expansion[2] (elaborating, extending or enhancing relations). These logical systems are found in many languages (see Caffarel *et al.* 2004). However, the structural organization which these systems are mapped onto for the construction of different kinds of logic differs across languages (see Matthiessen 2004: 575–80 for an illustration of such a difference between Akan and English in the distribution of labour between the experiential and logical modes of construing experience within the ideational metafunction; see also Martin 1995 for Tagalog). Raising learners' awareness of that difference is one good example for how conscious learning, 'learning through language', can foster successful FL/SL learning.

To explore that possibility I will examine clause complexing, first 'from below' with respect to its structural (syntagmatic) organization and then 'from roundabout' with respect to its systemic (paradigmatic) organization. Importantly, these two perspectives are complementary and offer learners opportunities for deploying a learning strategy that involves both comparison and contrast of their mother tongue and the target language.

Syntagmatic organization of language logic

While the dynamics of clause complexing across languages are multifaceted, they can nonetheless be generally characterized in terms of the two directions that linear development of the logical structure of a clause complex can take, namely progressive versus regressive (Martin 1995: 198). Progressive development represents a forward movement of logic and regressive development represents a backward movement. The major difference lies with the position of a 'nuclear' event in a nexus relative to other events that depend on it as 'satellite' or as sub-sequences, in order to form the total sequence of events that is realized as a clause complex (for the 'nucleus-satellite' relations in the rhetorical organization of text, see Mann and Thompson 1987, Mann *et al.* 1992; for clause complexing, see Matthiessen 2002 and Matthiessen and Thompson 1988).

In a clause nexus, there will thus be one nuclear clause and one satellite clause. The nuclear clause is an independent element in the nexus and is grammatically unconstrained; it has the full systemic potential characteristic of a free clause. For example, any terms from the interpersonal system of MOOD such as 'declarative' and 'imperative' can be selected freely (according to the speech functional categories they realize). The satellite clause, on the other hand, is a dependent element in the nexus and is grammatically constrained: only certain options within the full systemic potential are open to it. For example, a dependent clause cannot select for 'imperative' in the system of MOOD, nor can it enact politeness in terms of the system of POLITENESS.

The clauses in a nexus are related through 'parataxis' or 'hypotaxis', or 'coordination' and 'subordination' respectively in traditional grammar. In a paratactic nexus, the elements related are independent clauses that hold 'equal' status. In a hypotactic nexus, the elements related are of 'unequal' status and dependent elements are realized by dependent clauses. In this analysis, 'hypotaxis' includes only true subordination, not embedding, because embedding is addressed within the domain of group, where the embedded clause in Japanese serves as a modifier of the following head noun (see Matthiessen and Thompson 1988, for this point; see also Halliday and Matthiessen 2004: 382–3 for more detail about taxis in general). In Table 5.2, following Halliday (1994), paratactic structures are represented by numerical notation 1, 2, 3 . . .; hypotactic structures by the Greek letter notation α, β, γ, etc.; '^' indicates ordering, 'followed by'.

The difference between 'progressive' and 'regressive' linear development of a nexus is most evident in the hypotactic nexus, where the satellite element (e.g., β) depends on the nuclear element (e.g., α). In this respect, the logical organization of clause complexes in Japanese is distinctively different from that of English. Japanese is strictly regressive:[3] a nuclear element realized by an independent clause always follows a satellite element realized by a dependent clause in a hypotactic sequence, as in [β:] *ame ga huttara* [α:] *boku wa dekakenai* from Table 5.2. From an interpersonal perspective, the nucleus thus falls at the end of an interpersonal structure where the current speaker is just about to hand over the role of speaker to the current listener in discourse. By contrast, English can be either regressive or progressive, because the locus of a nucleus in a nexus can be either 'before' (i.e., progressive) or 'after' (i.e., regressive) its satellite element: *I won't go out if it rains* or *if it rains I won't go out.*

The distribution of semantic load that is shared between the satellite and the nucleus is also signalled differently among languages. In Japanese, logico-semantic relations such as time, cause and condition are signalled explicitly by structural conjunctions at the 'end' of the satellite-dependent clause, e.g., *ame ga hutte iru node* rather than at the 'beginning' as in English, e.g., *because it is raining.* That is, in Japanese the clause that initiates a move indicates its semantic relevance at the end of the initiating clause and does so

Table 5.2 Parataxis and hypotaxis

	secondary clause	primary clause
(a) parataxis [1 ^ 2]	[1:] *Ame ga* [Process:] *huru kara* rain GA fell because 'Because it will rain'	[2:] dekakeru na 'don't go out'
(b) hypotaxis [β ^ α]	[β:] *Ame ga* [Process:] *huttara* rain GA fell-conditional 'If rain falls'	[α:] boku wa dekakenai 'I won't go out'

irrespective of taxis (see Table 5.2). To highlight that structural difference, I will follow the notational conventions proposed by Halliday (1994): notations that indicate interdependency relations (e.g., '×' for 'enhancement') are inserted where structural conjunctions appear, i.e., for Japanese this occurs 'after' the tactic markers, e.g., $\beta^{\times}$ *ame ga hutte iru node*, for English 'before', e.g., $^{\times}\beta$ *because it is raining*.

In the example in Table 5.3 the original English sentence comes from the UTS-Macquarie Corpus of spoken Australian English; the Japanese sentence is a translation equivalent.

Table 5.3 Tactic organization in English and Japanese

(a)	$^{\times}\alpha\alpha$ ^	$\alpha\beta$ ^	$^{\times}\beta$
	We would all be dead,	'cause there wouldn't be oxygen,	if there weren't trees on the earth.
(b)	$\beta\beta^{\times}$ ^	$\beta\alpha^{\times}$ ^	-
	Moshi chikyuu ni ki ga nakunattara,	sanso ga nakunatte,	min'na shinde shimau da daroo
	'if there weren't trees on the earth'	'oxygen wouldn't be there'	'we would all be dead'

Using arrows to indicate the direction of the development (i.e., progressive '→' and regressive '←'), the linear development of clause complexes in the above examples can be made even more explicit.

Table 5.4 illustrates the fundamental difference between Japanese and English in the way logic is constructed serially and recursively.[4] Because structure-based SL/FL teaching often fails to deal with the inherent property of 'recursion' for the construction of logic, even advanced learners tend to be unaware of this important fact about languages and get confused when they meet more intricate clause complexes. An extract from a student's learning journal, exemplifies the challenges this poses in learning Japanese:

> The whole thing was quite complicated but I think it did help and I got some important points out of it. Firstly, in English, there is both regressive and progressive logic organization. However, in Japanese, everything is regressive which is why you have to listen to the end in order to understand what someone is saying. I found that was a really interesting point and it is so true! I don't think I was very conscious of it before, but now I am.

Paradigmatic organization of the language logic and learning strategies: comparing and constrasting

The logical structures discussed above are examples of a univariate structure (see Halliday 1965) that serves to realize terms in logical systems in general and in the system of taxis in clause complexing in particular (cf. Halliday 1979). These univariate structures combine clauses into clause

Table 5.4 The dynamic movement of regressive and progressive logic

	internal nesting and interdependency			
	regressive		regressive + progressive	progressive
English:			${}^{\times}\beta \leftarrow \alpha\alpha \rightarrow \alpha^{\times}\beta$	$\alpha\alpha \rightarrow \alpha\beta \rightarrow \beta$
Japanese:	$\beta\beta^{\times} \leftarrow \beta^{\times}\alpha \leftarrow \alpha$	$\beta\beta^{\times} \leftarrow \alpha, \leftarrow \beta^{\times}\alpha$	–	–

nexuses, and clause nexuses into clause complexes. While a clause complex can thus be fairly described as a logical complex of clauses, it is important to recall that clauses are multifunctional constructs. A clause is simultaneously a message (a quantum of information in the flow of information: textual), a move (a quantum of dialogic interaction: interpersonal) and a figure (a quantum of change in the flow of events: experiential). As such it represents a grammatical unification of these three metafunctional units of meaning – message, move and figure (see Halliday and Matthiessen 1999, 2004: 588–92). In other words, as 'a sequence of messages, of moves, and figures' (Matthiessen 2002: 259) a clause complex serves as an important textual, interpersonal and ideational domain in the creation of meanings in text. Its logical meaning is derived from 'metafunctional unification', the unification of different strands of meanings that are mapped globally onto a univariate serial structure of clause complexes by the grammar.

In what follows, I will illustrate some of the patterns of metafunctional unification that are engendered in the realization of logical meanings in clause complexes. The approach adopted is based on the trinocular perspective proposed by Halliday (e.g., Halliday and Matthiessen 2004: 31, 119), which enables inquiry from different but complementary angles. We can view clause complexing 'from above' – from the vantage point of rhetorical patterns in text at the level of semantics; 'from below' – from the vantage point of structural marking of logical linkage; and 'from around' – from the vantage point of the grammatical systems of taxis and logico-semantic type that operate in clause complexing. Significantly, the trinocular approach enables us not only to capture in a holistic way how natural logic is construed by clause complexing but also to identify the pitfalls that are concealed in a structure-biased approach to teaching and learning clause combining or in structure-biased language teaching in general. By contrast, the trinocular perspective adopted here is analogous to the general learning strategies of comparing and contrasting. For Halliday (1999: 72) ' "comparing" means finding likenesses among things that are different, "contrasting" means finding likenesses among things that are alike'. As the student learning journals indicate, these strategies are central to SL/FL learners' learning through language; ultimately, they are strategies for learning to think grammatically.

Specifically, in the context of SL/FL intermediate to advanced learning, 'comparing' generally takes place where grammatical thinking arises 'from above', that is, where similar logico-semantic relations, such as relations of temporal progression, are expressed through different structural conjunctions, as in [*shi*]-*tara* 'when' and [*suru*] *to* 'and then'; contrasting occurs 'from below', where the same expression, such as the conjunction *ga* ('but'), can realize such different logico-semantic relations as adversative versus introductive. In the process of comparing and contrasting, metafunctional realization of logical meanings plays a key role. The typological variables illustrated below serve as an analytical tool for enabling learners to distinguish between the likeness among things that are different and the likeness among things that are alike.

Realization of subjective and objective logic

Broadly speaking, Japanese operates with two general orientations in logic: one is where logic is 'interpersonalized' (or 'subjective') and the other is where it is 'experientialized' (or 'objectified') (cf. Matthiessen this volume for the notions of 'subjective' and 'objective' representation of places of interest in English and German guidebook texts; cf. also Halliday and Hasan 1976 and Martin 1992 on internal versus external conjunctive relations). Below each type is discussed in turn.

(1) When the logic is 'interpersonalized', the logical connection between the two clauses of a clause nexus is based on their nature as interpersonal moves and involves the speaker's subjective viewpoint: it is about 'enactment' of personal experiences in a local context, e.g., [$1^{\times}$:] *Atama ga itai <u>kara</u>*, [2:] *kaette mo ii desu ka?* 'I have a headache, *so* is it OK to go home?'. This type of clause complex typically ends with a clause representing a move that demands information or goods and services from the listener. In general, such a clause is in the interrogative or imperative mood or else in the declarative mood but marked in terms of other interpersonal systems such as modality.

(2) When the logic is 'experientialized', the logical connection between the two clauses in a nexus is based on their nature as experiential figures: the nexus construes a link between events in the world of our experience and 'theorizes' that experience rather than regulating behaviour (Matthiessen 1993, Okuda 1986), e.g. *hairu koto wa dekiru <u>ga</u>, deru koto wa dekinai* 'we can enter *but* cannot get out'. This type of clause complex typically provides information in the form of a 'statement' realized by declarative mood.

The difference between these two types is multifaceted and thus cannot be dealt with in full detail here. However, the following example of one of the most common structural conjunctions, *ga* ('but'), illustrates their fundamental differences, the very differences that scaffold learners' grammatical thinking.

Table 5.5 Structural conjunction *ga*: adversative and introductive type

	structure	relation	secondary clause	primary clause
(a)	1+ ^ 2	adversative	Mada haru da ga, 'It is still spring, but'	totemo atsui 'it's very hot'
(b)	1= ^ 2	introductive	Samui deshoo ga, 'It may be cold, but'	doa o akete kudasai. 'please open the door'

The use of the structural conjunction *ga* ('but') seemingly makes sentences similar. This is the view of the clause nexus obtained 'from below', which focuses on the deployment of *ga*. The view 'from above', from the vantage point of semantics, and the view 'from around' with respect to grammar, on the other hand, give us a different picture of their likeness and/or difference. That is, while the use of *ga* is the same in (a) and (b), these clauses are linked by relations of different kinds and instantiate different grammatical environments.

Figure 5.1 below illustrates the contrast between the two types of logical relation that are realized by the structural conjunction *ga* ('but') in the systemic environment of 'extension' and 'elaboration'.

In Example (a), the primary clause expands the secondary clause, 'extending' it by adding the primary event that is adverse to, or contrary to, what is normally expected in view of the secondary event. Hence, the logico-semantic relation is 'adversative' to the preceding *ga*-marked clause in the primary clause. Grammatically, an adversative clause complex always ends in the declarative mood. That is, the speaker presents a piece of adversative reality by turning it into a statement; the logical link is experiential in orientation, connecting events in our experience of the world.

In Example (b) on the other hand, the secondary clause serves as a kind of preface, based on which the speaker makes a move to demand information or goods and services from the listener. This is the 'introductive' type (one subtype of 'expository' among various elaborating types), where the introductory comment or thesis presented in the secondary clause is exemplified, clarified or explicated further in the primary clause, the clause that demands a response from the listener. In terms of MOOD, the primary clause is generally interrogative (the speech function of 'question') or optative mood (that of 'desire'); if it is declarative, the clause carries a subjective overtone by accompanying negotiatory particles such as *ne* and *yone*, based on which the speaker requests from the listener comments, evaluations or the like (see Figure 5.5 where the Predicator presents an explanation by *n da* '(the reason for that is'). This logic is interpersonalized according to the speaker's subjective view of social life in a given communicative context.

The advanced Japanese language learner's account of the topic, taken from her learning journal, shows this kind of grammatical thinking taking

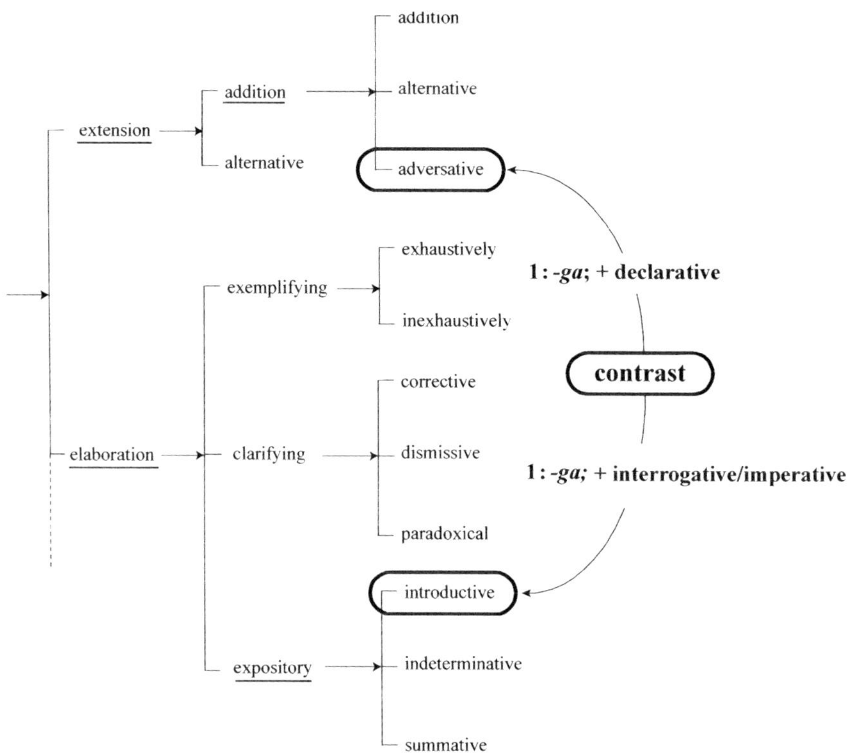

Figure 5.1 Contrast between the adversative and introductive types: 'from below'

place through the process of contrast, i.e., viewing the domain from below with respect to the overt structural marker, a view that is complemented by a view from above with respect to the systemic environment in which the clauses are combined into a clause complex.

> Expository is made up of three subtypes being introductive, indeterminative, and summative. The introductive type is marked by *ga* and *keredomo*, which is the same as the adversative type for extension. However, they differ in the sense that the extension type can only have a declarative clause after the conjunction. Whilst the elaboration type will have an interrogative clause, optative clause or declarative clause with Negotiator such as *ne* and *yone* at the end. I can easily remember this type as the preceding clause acts in a way to suggest there is more coming, as it is more of an opening comment and thus in this way elaboration occurs when the following clauses are realized.

The systemic contrast between the two types is engendered by the logical metafunction. However, the illustration makes two important points: first, to show that human logic is created in response to social participation in and making sense of our experiences of the world that has been evolving to the current socio-economic and technological state; and, second, to show that the creation of logical meanings that is metafunctional in origin is reflected in the covert (semantic) operation of the grammar.

The following example adopts the strategy of comparison by viewing the domain of temporal sequence 'from above', identifying ways of construing temporal order of events in reference to two structural conjunctions, [*suru*] *to* 'if, when' and [*shi*] *tara* 'supposing that', whose roles are often identified solely with the meaning of conditionality (or else no meaning is discriminated, particularly at the introductory level). Contrary to the previous set of examples, the structural difference here signifies likeness in meaning. Consider the following examples in Table 5.6:

In Examples (a) and (b), two events are combined through verbs in different conditional forms that serve as structural conjunctions: 'anticipated' form [*suru*] *to* and 'suppositional' form [*shi*] *tara*. As already stated, these forms are often defined as an expression of 'conditionality'. However, the meaning they assign to the clause complex in the given clausal environments is not one of condition, but rather one of 'temporal sequence', in particular, 'temporal immediacy', in that the primary event is brought into being by chance in immediate succession to occurrence of the secondary event. One probe that can be applied to the temporal interpretation of these events pertains to their metafunctional potential: the meaning of 'immediacy' could be augmented 'experientially' by a circumstance of time in the primary clause, e.g., *sugu* 'immediately', *totan ni* 'just then', as in (a) . . . *sugu denwa ga natta* 'the phone rang *immediately*', and (b) . . . *totan ni mune ga kurushiku natta* '*suddenly* (I felt) oppressed in the chest'.

If, on the other hand, the meaning is one of conditionality, this experiential potential is sealed off from being selected and the interpersonal meaning potential becomes available to support the conditional interpretation. That is, other things being equal, if events were of a conditional relation, they should allow for mood Adjunct *moshi*[*mo*] 'if' to enter into the organization and enhance the suppositional nature of a given proposition/proposal, e.g.,

Table 5.6 Temporal immediacy realized by different structural conjunctions

	structure	type of relation	secondary clause	primary clause
(a)	β× ^ α	temporal: immediacy	Ie ni hairu-to 'when I entered the house'	denwa ga natta. 'the phone rang'
(b)	β× ^ α		Heya ni hait-tara 'when I entered the room'	mune ga kurushiku natta. 'I felt oppressed in the chest'

(a) *moshi ie ni hairuto* . . . 'if you enter the room . . .'.[5] In other words, the meaning of temporality or conditionality that is signalled by the same structural conjunctions can be identified by reference to the metafunctional potential. Figure 5.2 illustrates this in the systemic environment of enhancement.

What, then, is the difference in the 'likeness in meaning' realized by these different wordings? The answer lies in the general types of logic, as either experiential (objective) or interpersonal (subjective) in orientation. That is, Clause (a) with the 'anticipated' form [*suru*]*to* belongs to the former type and Clause (b) with the 'suppositional' form [*tara*] to the latter. With (b), the speaker may exchange his/her logic by turning the primary event into an enactment of a command that is realized by the imperative form, whereas this is not possible with clause (a). In other words, the two different ways of representing the logic of temporality or conditionality occur in different social contexts, depending in this case on the interpersonal role that language logic is made to play.

Textual aspects of logic and their role in guiding into the rhetoric of text

In the previous section, we examined metafunctional unification in the domain of clause complexing, in particular with reference to the interpersonal system of MOOD and its realizational relationship to a set of structural conjunctions. We also observed how two general learning strategies of comparing and contrasting can be effectively deployed to view clause complexes from two complementary angles: 'from below' with respect to expression and 'from above' with respect to meaning. In this section, we will extend our discussion by observing the contribution both experiential and textual functions make towards clause complexing. This leads us from the domain of clause complexes into that of the rhetoric of discourse.

Experiential theme in intra-clause relationships

In the following clause complex consisting of four clauses our concern is with textual and experiential 'intra-clause' relationship of clause complexes, that is, with how each clause is combined into a clause complex by textual and experiential functions. To complement the view obtained this way, the next step examines the 'inter-clause' relationship of the clause complex to its discourse environment.

Figure 5.3 below illustrates two grammatical phenomena at work within a clause complex, i.e., thematic continuation of Theme that runs across the entire complex (indicated by arrows), and Subject/Medium identification in the hypotactic nexus of two clauses (indicated by lines with dots at their ends).

In Japanese, the presence or absence of Subject and/or Theme in the clause (complex) is subject to its semantic continuity in a given grammatical environment (Teruya 2004, in press): as long as there is a semantic continuity

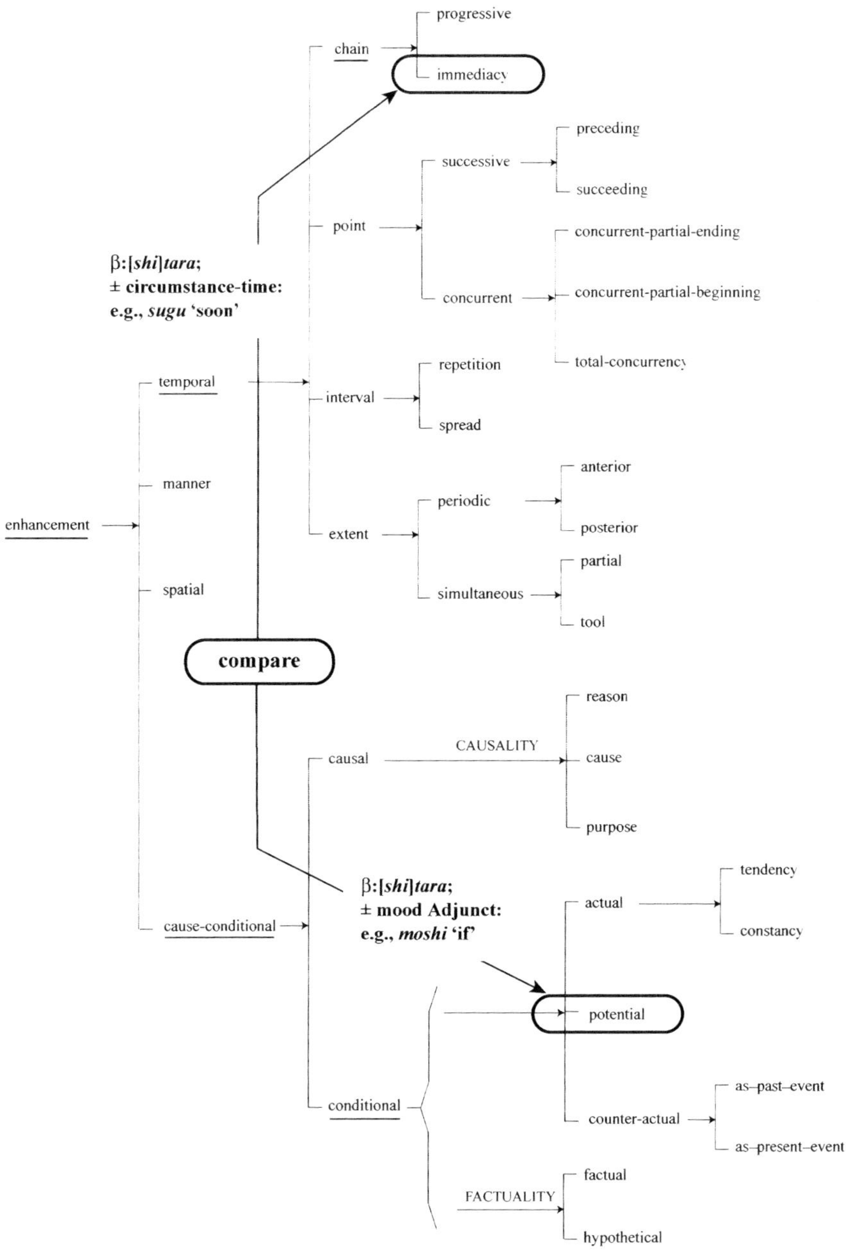

Figure 5.2 Comparison between temporal immediacy and conditional potential: 'from above'

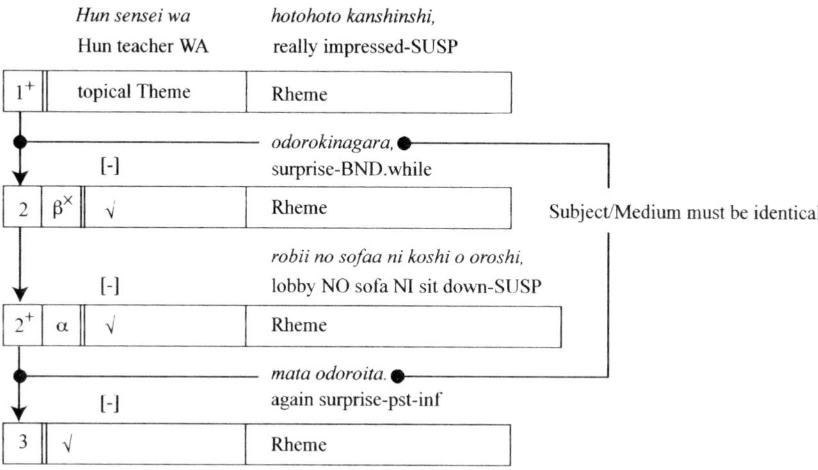

'Dr. Hun was really impressed, while still surprised,
sat down on the sofa at the lobby, and again got surprised.'

Figure 5.3 Semantic continuity of Subject and/or Theme in the clause complex

of Subject and/or Theme from previous discourse or else as long as it is inferrable from a given context, Subject and/or Theme tend to remain implicit until such time when it is reset by another explicit Subject and/or Theme. This textual and interpersonal generalization about the functional behaviour of Subject and Theme guides learners to identify not only how a series of clauses in a nexus is semantically, thus cohesively, organized but also how to recover an implicit Subject in order, for example, to translate the clause complex into English, where the Subject almost always figures in when the complex indicates indicative mood (see Halliday 1994, Halliday and Matthiessen 2004).

In terms of the tactic organization, the example has only one hypotactic clause that is chained into a sequence of P ^ H ^ P ^ P (P= paratactic, H=hypotactic relation), as signalled by each clausal ending (*-shi* 'and' = paratactic; and *–nagara* 'while' = hypotactic); there is one explicit Theme conflated with the Subject figured at the beginning of the complex. Following the previously stated general tendencies students were able to reason grammatically in their learning journals about the boundaries of clauses in the complex, correctly discerning that there may not be a single line of dependency when there are several clauses:

> . . . when you have a hypotactic clause followed by a paratactic clause, the hypotactic clause is bound to the paratactic one, therefore you must set up a boundary after the paratactic clause.

Accordingly, the interdependency sequence is now analysed as P ^ (H ^ P) ^ P (brackets indicate dependency). These boundaries are semantic in nature in that they are derived from the interdependency relations reflected in the verbal forms. More important, since the boundary is semantic, clause boundaries are metafunctional in their makeup. The same student's subsequent observation halfway through the 14-week course addresses matters experientially:

> . . . I'm not sure if I understood this correctly but I am still of the belief that one cannot just rely on the interdependency relations to organize the clauses logically as you still have to understand the meaning of it before you can understand the logic.

The point is valid and requires explanation with respect to the grammar. Linking two clauses into the nexus (H ^ P) by the hypotactic structural conjunction *nagara* 'while' is motivated experientially. That is, when two clauses are linked in this way, the nexus is concerned with the representation of 'concurring events that are compounded at one time', rather than being sequentially organized in time. For this experiential nature, the linked two clauses share the same Agent/Medium through which the concurrence of two events is brought into being by the identical participant, i.e., 'while [he] was still surprised, [he] sat down on the sofa at the lobby'.

Interpersonally, on the other hand, the fact that the Medium 'Hun teacher' is implicit in the dependent clause points out that the Subjects of the two clauses in the nexus are also identical. The principle is this: if the Subject is different in the dependent clause it is made explicit and always signalled by a postpositional marker '*ga*'. Another quote from a student's learning journal refers to that realization:

> . . . another way of analysing the clause structure was to look at what the subject is in each clause; clauses with the same subject tend to be in the same subgroup.

Textually, these experiential and interpersonal features are related to the overall discourse theme through the textual function of Theme. Once a Theme is introduced, it remains implicit until a new Theme resets the discourse theme. In other words, while explicit Theme records the highest value of thematic amplitude, the textual implicitness of Theme signifies not only the scope of semantic continuity of clauses for the same Theme (Teruya 2004, Hinds 1983) but also, in the case of clause complexes, co-referentiality between the Subject in the dependent clause and the Subject/Theme of the independent clause, or the nexus as a whole (cf. Matthiessen 2004: 637 for 'switch reference system').

Textual Theme in discourse environment

Discussion of Theme has thus far addressed experiential Theme, one of the three types of Theme (the others being logical and interpersonal) that are similarly slotted in at the clause-initial position that holds thematic significance; it is a topical Theme that is served by participants in the experiential configuration. Continuing our consideration of the inter-clause relationship of logical function with regard to the rhetorical organization of discourse, we now shift our perspective from clause (complex) to text and observe textual Theme that is realized by cohesive (non-structural) conjunctions, such as *daga* 'but'. The shift of perspective allows us to capture the relationship between cohesive conjunctions serving as Theme and structural conjunctions. In turn, this demonstrates how the kind of grammatical thinking illustrated for clause complexing can be further extended and related to thinking about the rhetorical organization of text.

The perspective is heuristically effective because in Japanese there are strong parallels between structural conjunctions and cohesive conjunctions in terms of their morphological realizations (see Martin 1992, Chapter 4 for the rhetorical nature of conjunctions in English). In SL/FL contexts, that realizational similarity enables learners to engage in the process of contrasting between clause (complex) and discourse by approaching it 'from below', moving from the similarity in wording between these two types of conjunctions, to similarity and difference in meaning that they create in the environment of clause complexes and discourse.

Below is an extract from a radio interview by Japanese Prime Minister Koizumi used in the course. Two clauses in the extract are analysed rhetorically in order to illustrate the external relation between clauses (1) and (2) (for the notion of 'external relation', see Martin, *ibid.*). The relevant cohesive conjunction is *soo shitara* 'then/if so'. In general, conjunctions are realized clause-initially and hold textual significance as a logical Theme. In the example, the conjunction appears at the thematic position of Clause (2), setting up a rhetorical relation between what has gone before and what is to come next in the discourse. It is derivatively related to the structural conjunction [*shi*] *tara* ('suppose that') already observed in the previous section.

As discussion of the environment of the clause complex revealed, the logical meaning that the structural conjunction brings about depends on other functional features embodied in the complex: for example, the meaning of 'temporality' was connected to the experiential potential of the primary clause for having a circumstance of time that defines the immediacy of the two events in the nexus, while 'conditionality' is related to the interpersonal presumption of a given event that is enacted through a mood Adjunct such as 'if'.

The functional organization of the rhetoric of the text in which this conjunction *soo shitara* 'then/if so' occurs is very similar to that described for its structural counterpart. The same is true for the meanings associated with it: 'then' as temporality and 'if so' as conditionality. As illustrated in Figure 5.4, cohesion is established based on the sequence of (1) and (2) that occurs within the temporal framework that is defined by the underlined circumstance of temporal extent *kono hantoshi de* 'within half a year' in Clause (2).

Note here that the parallelism between the conjunction and its structural coupling is by no means the sole cohesive relation that may be construed in the given discourse environment. The discourse environment may govern a cohesive interpretation in a different way. For example, while the sequence of satellite (1) and nuclear (2) is in fact one of temporal relations, it is at the same time one of causal relation, as reflected in the English translation given by an advanced learner in Figure 5.4: (1) invokes the occurrence of (2) in the overall scheme of things that the prime minister is engaging in his radio interview. Discovery of such a parallelism is a very important strategy that learners can employ as a kind of scaffolding in order to carry out the process of 'contrast'.

(Text 5.1 Radio interview)
Interviewer: Soo datta n desu ka. ||| Shirimasen deshita. |||
Oh really? I had no idea.

Hosokawa: Kaisha mo kantan ni tsukureru yoo ni natta n desu yo. ||| Ima made wa kabushikigaisha o tsukuru no ni shihonkin to shite issenman en hitsuyoo datta n desu. ||| (1) Kotoshi no nigatsu kara shihonkin ichi en demo kaisha o tsukureru yooni shita n desu. ||| (2) Soo shitara, kono hantoshi de yonsen o koeru atarashii kaisha ga tanjoo shimashita. ||| (3) Kisei kaikaku o susumereba, || zeikin o toonyu shinakutemo, || minkan no yaru ki to aidea ga ikasarete || atarashii koto ga dekiru yooni naru n desu. |||

'It has become possible for companies to make these things easily too. Until now, in order to create a joint stock corporation, it was necessary to pay 10 million yen. (1) From February this year, it is possible to start a company with a capital of even as little as 1 yen. (2) Because of this, within half a year, new companies numbering over 4,000 have come into being. (3) If we continue this readjustment reform, without investing tax funds, the public's motivation and ideas come to life and new ways of conducting businesses become possible.'

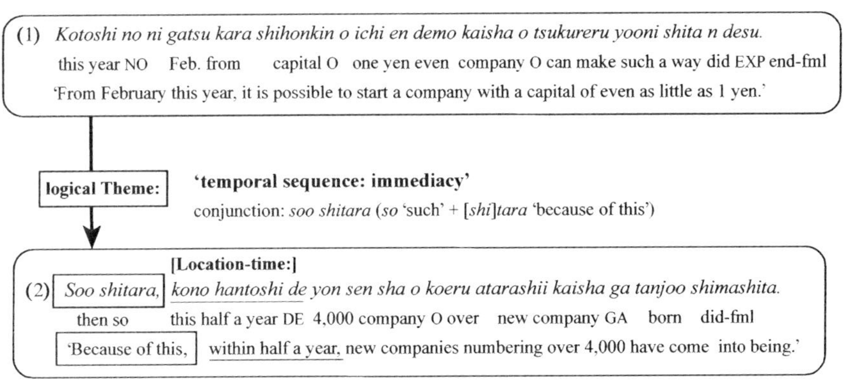

Figure 5.4 Cohesive conjunction and its external functional environment

Metafunctional unification through global mapping of the metafunctional maps

I have discussed the contribution that the metafunction systems of SUBJECT, THEME and PROCESS TYPE each make towards the construction of logical meaning. The mechanism behind the metafunction unification that creates logical meaning is what I call 'a global mapping' in the construction of language logic, a term I borrow from Edelman (1992: 89–91, 101–5). However, there is an important difference. Edelman uses the term 'global mapping' to refer to the dynamic structural mapping of psychology on to physiology for the emergence of perceptual categorization, e.g., 'brain-based memory' that 'results from a process of recategorization' (*ibid.*: 102), through re-entry by which functionally segregated maps are linked in time through 'parallel selection and the correlation of the maps' neuronal groups, which independently and disjunctively receive inputs' (*ibid.*: 84). By contrast, my use of the term is intended to capture its semiotic operation, the dynamic mapping of the stratified maps of meaning and expression in time through metafunctional unification that recursively correlates and coordinates 'intrinsic' experiential and interpersonal meanings selected systemically and integrated textually in a set of clauses in constructing the 'extrinsic' logical meaning, which is synthesized dynamically in the serial structure of clause complexes.

That is, the 'language-based logic' is engendered extrinsically through this kind of global mapping, which interconnects the functionally segregated interpersonal, experiential and textual metafunctions that form the various patterns of their couplings that 'loop back' on themselves serially. This system's property of 'looping back', which forms lexicogrammatically the logical system of 'recursion', is what makes the logical systems dynamically complex (see Lemke 1995: 108 for the nature of complex systems). The ability to create this complex dynamic of recursive movements of language in speaking and writing and to trace it back in listening and reading critically distinguishes intermediate and advanced SL/FL learners from those at lower levels.

Figure 5.5 below illustrates such a dynamic global mapping in the clause complexing of three clauses combined into a nexus. Notations used are as follows: three sets of three square boxes that are internally connected by solid lines represent metafunctional clausal makeup: top (textual), below left (experiential) and below right (interpersonal); one larger square represents a nuclear-satellite relation framed by hypotactic relation. Solid lines indicate the intra-clause relationship that is organized within each metafunction (here the experiential configuration is organized around the process and the interpersonal prosody around the predicator (see Halliday and Matthiessen 1999: 52–8 for 'frame-representation' in computational context). Dotted lines represent inter-clause relationships that, together with intra-clause relationships, form a global mapping that is taking place in the given example. Letters represent functions: T=Theme, R=Rheme; S=Sensor, A=Actor,

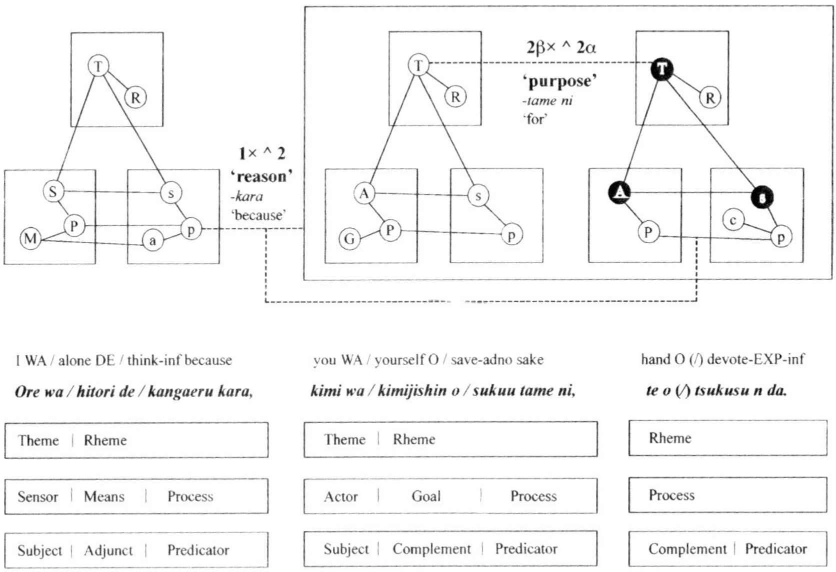

'I will think by myself, so you do everything you can to save yourself.'

Figure 5.5 'Global mapping' engendering logical meaning in clause complexing

M=Means, G=Goal, P=Process; s=Subject, c=Complement, a=Adjunct, p=Predicator; white background indicates that they are explicit, black background that they are implicit in the clause complex.

Conclusion

In this chapter, I have proposed that grammatical thinking (or learning through language) is a particularly effective learning strategy that advanced learners can adopt to expand their meaning potential, for instance, in the construction of the natural logic embodied in the grammar of clause complexing in Japanese and other languages. I have supported this claim by illustrating metafunctional unification and by taking examples from learning journals. The grammatical thinking illustrated here is supported by a systemic-functional account of the grammar of Japanese which, in the context of SL/FL learning, offers linguistic scaffoldings that enable learners to compare and contrast grammatical selections that are mapped on to the dynamic structure of clause complexes. I have also pointed out that, by decoding an example of the grammatical mechanism at work behind the meaning of language logic embodied in the complex, learners can come to appreciate the importance of systems thinking in SL/FL learning. This is crucial because one way to become an advanced SL/FL learner is through awareness of the system's inherent property of 'recursion' or 'looping back' that brings about open-ended complex structures. In current SL/FL Japanese language teaching and learning, this grammatical property has largely been neglected or gone unnoticed, in spite of its importance in many types of discourse.

Within this logical systems property of recursion, I have identified a grammatical mechanism that I called a global mapping in semiosis, in order to capture its power of creating language logic. In the environment of language, this global mapping is semiotic in nature. For that reason it enables the mapping of varying metafunctional meaning-making systems to create grammatical couplings according to which learners can rearrange, collapse or replace another such mapping that is being built in the course of expanding one's meaning potential in the target language.

The kind of grammatical thinking that involves conscious learning of language through language systems is not just useful in learning the grammar of the language but also critical in learning 'about' the way language functions in the society within which the language operates. With reference to the extract taken from the radio interview in which Koizumi addresses his general audience (Text 5.1, see Clause (3)), learners 'discovered' relationships between his way of using language and the socio-political purpose of his discourse, describing it as follows:

> Also interesting was the way that politicians use hypotactic clauses in order to make what they are saying unarguable.

A second characterization states:

> We were also given a sample of the Prime Minister Koizumi's speech and Sensei (i.e., teacher) remarked that all his political views are in hypotaxis. He does this because hypotaxis are bound clauses and therefore no-one can interrupt. However, when the Prime Minister is talking to his people, he uses parataxis so it is easier for people to understand. I also thought the logic behind using grammar in this way was really remarkable and something I had never heard of before.

Such discovery on the part of learners in the course of self-learning can provide a crucial counterweight to a kind of SL/FL learning that limits engagement with grammar to learning rules that govern pre-defined, static structures and thus fails to capture the dynamic nature of clause complexing. Indeed, one way to describe the successful transition from intermediate to advanced stages of SL/FL learning is in terms of a movement that provides to learners linguistic resources that are 'functional' in the sense that they cannot only identify functions that are inherently internal to language itself but realize as well the social functions that the language serves for various communicative purposes.

Notes

1 In transcribing Japanese, a modified Hepburn style is used. Modifications made to the style include: long vowels are expressed by repeating a vowel, as in *aa*, *ii*, but for proper names no distinction between short and long vowels is made; [ɸɯ] is written /hu/ instead of /fu/. In the interlinear glossing given in the chapter, nominal markers such as GA and O are indicated in upper case (see Teruya 2004: 187–8); morphological information that helps identify particular functions is glossed as follows: BND: binder; fml: formal; EXP: explanative mood; inf: informal; SUSP: suspensive form.

2 To save space, the following discussion addresses the expansion type only. However, metafunctional unification of the kind described in the chapter is applicable for projection as well (see Teruya 2004, in press).

3 This is true at the rank of clause and in clause combining only. At the rank of group, e.g., the verbal group, the development of interpersonal logic is progressive: experiential content of doing, sensing, saying and being is augmented progressively by various interpersonal meanings, such as modality and evidentiality (see Teruya, in press).

4 In the example one of the secondary clauses and the primary clause are inverted; as a result the secondary clause comes to take up the final clausal position. Intonationally, the secondary clause that is inverted this way is pronounced with a falling pitch accompanied by decreased loudness; together this indicates its secondary, or additional, nature to what precedes it (Uemura 1989: 213).

5 In reality, mood Adjunct *moshi* 'if' cannot enter into these clause complexes because the expression of a suppositional condition is applicable to the event that has the 'potential' to occur or else is 'counter-factual' to what has already happened (see Teruya, in press). The examples are past events, thus are not subject to an interpersonal presumption.

References

Bybee, J. and Noonan, M. (eds) (2002) *Complex Sentences in Grammar and Discourse: Essays in Honor of Sandra A. Thompson*. Amsterdam/Philadelphia: John Benjamins.

Caffarel, A., Martin, J. R. and Matthiessen, C. M. I. M. (eds) (2004) *Language Typology: A Functional Perspective*. Amsterdam/Philadelphia: John Benjamins.

Edelman, G. M. (1992) 'Bright Air, Brilliant Fire: On the Matter of the Mind'. New York: Basic Books.

Haiman, J. and Thompson, S. A. (eds) (1988) *Clause Combining in Grammar and Discourse*. Amsterdam: John Benjamins.

Halliday, M. A. K. (1965) 'Types of structure. The O.S.T.I. Programme in the Linguistic Properties of Scientific English'. Reprinted in M. A. K. Halliday and J. R. Martin (eds) (1981), *Readings in Systemic Linguistics*. London: Batsford, pp. 29–41.

Halliday, M. A. K. (1978) *Language as Social Semiotic: The Social Interpretation of Language and Meaning*. London: Edward Arnold.

Halliday, M. A. K. (1979) 'Modes of meaning and modes of expression: types of grammatical structure and their determination by different semantic functions', in D. J. Allerton, E. Carney and D. Holdcroft (eds), *Function and Context in Linguistic Analysis: A Festschrift for William Haas*. Cambridge: Cambridge University Press, pp. 57–79.

Halliday, M. A. K. (1994) *An Introduction to Functional Grammar*. 2nd edn. London: Edward Arnold.

Halliday, M. A. K. (1996) 'Literacy and linguistics: a functional perspective', in R. Hasan and G. Williams (eds), *Literacy in Society*. London: Longman, pp. 339–76.

Halliday, M. A. K. (1999) 'Grammar and the construction of educational knowledge'. The International Conference 'Language Analysis and Description: applications in language teaching'. Lingnan College and The Hong Kong University of Science and Technology, 26–29 June 1996, in R. Berry, B. Asker, K. Hyland and M. Lam (eds) *Language Analysis, Description and Pedagogy*. Hong Kong: Language Centre, The Hong Kong University of Science & Technology and Department of English, Lingnan University, pp. 70–87.

Halliday, M. A. K. (2000) 'Computing meaning: some reflections on past experience and present prospects', in H. Guowen and Z. Wang (eds), *Discourse and Language Functions*. Beijing: Foreign Language Teaching and Research Press, pp. 3–25.

Halliday, M. A. K. (2002) 'On grammar and grammatics', in R. Hasan, C. Cloran and D. Butt (eds), *Functional Descriptions: Theory in Practice*. Amsterdam/Philadelphia: John Benjamins, pp. 1–38.

Halliday, M. A. K. and Hasan, R. (1976) *Cohesion in English*. London: Longman.

Halliday, M. A. K. and Martin, J. R. (1993) *Writing Science: Literacy and Discursive Power*. London: Falmer Press.

Halliday, M. A. K. and Matthiessen, C. M. I. M. (1999) *Construing Experience through Meaning: A Language-based Approach to Cognition*. London: Cassell.

Halliday, M. A. K. and Matthiessen, C. M. I. M. (2004) *An Introduction to Functional Grammar*. 3rd edn. London: Edward Arnold.

Hinds, J. (1983) 'Topic continuity in Japanese', in T. Givón (ed.), *Topic Continuity in Discourse: A Quantitative Cross-language Study*. Amsterdam/Philadelphia: John Benjamins, pp. 44–93.

Lemke, J. L. (1995) *Textual Politics: Discourse and Social Dynamics*. London: Taylor & Francis.

Mann, W. C. and Thompson S. A. (1987) *Rhetorical Structure Theory: A Framework for the Analysis of Texts*. Marina del Rey, CA: USC/Information Sciences Institute.

Mann, W. C., Matthiessen, C. M. I. M. and Thompson, S. A. (1992) 'Rhetorical structure theory and text analysis', in W. C. Mann and S. A. Thompson (eds), *Discourse Description: Diverse Linguistic Analyses of a Fund-raising Text*. Amsterdam/Philadelphia: John Benjamins, pp. 39–78.

Martin, J. R. (1992) *English Text: System and Structure*. Amsterdam/Philadelphia: John Benjamins.

Martin, J. R. (1995) 'Logical meaning, interdependency and the linking particle -ng/na in Tagalog'. *Functions of Language*, 2, 189–228.

Matthiessen, C. M. I. M. (1993) 'The object of study in cognitive science in relation to its construal and enactment in language'. *Cultural Dynamics*, 6, 187–242.

Matthiessen, C. M. I. M. (1995) *Lexicogrammatical Cartography: English Systems*. Tokyo: International Language Sciences Publishers.

Matthiessen, C. M. I. M. (2002) 'Combining clauses into clause complexes', in J. L. Bybee and M. Noonan (eds), pp. 235–319.

Matthiessen, C. M. I. M. (2004) 'Descriptive motifs and generalizations', in A. Caffarel, J. R. Martin and C. M. I. M. Matthiessen (eds), pp. 537–674.

Matthiessen, C. M. I. M. and Thompson, S. A. (1988) 'The structure of discourse and "subordination" ', in J. Haiman and S. A. Thompson (eds), pp. 275–330.

Okuda, Y. (1986) 'Jookenzuke o hyoogensuru tsukisoiawase-bun: sono taikeisei o megutte (Dependent clauses expressing conditions: on their systemicity)'. *Kyoiku Kokugo* 87 (*Educational Japanese* 87), 2–19.

Teruya, K. (2004) 'Metafunctional grammar of Japanese', in A. Caffarel, J. R. Martin and C. M. I. M. Matthiessen (eds), pp. 185–254.

Teruya, K. (in press) *A Systemic Functional Grammar of Japanese.* 2 volumes. London: Continuum.

Uemura, Y. (1989) *Nihongo no intoneeshon (Intonation in Japanese).* Kotoba no kagaku 2 (Science of Language 3). Gengogaku Kenkyukai (Linguistics Working Group). Tokyo: Mugi Publishers, pp. 193–220.

6 The linguistic features of advanced language use: the grammar of exposition

Mary J. Schleppegrell

Introduction

Enabling students to develop advanced language capacities is a key goal both for foreign language education and for the education of students learning English as a second language in school settings. This chapter examines descriptions of advanced language use and suggests that we can expand our understanding of this construct by incorporating a functional linguistics perspective, linking particular tasks with the language choices that most effectively realize those tasks. It uses the expository writing of high school English learners to illustrate the linguistic challenges of advanced literacy in this context, identifying grammatical features that contribute to the construction of texts considered 'advanced'.

For foreign language instruction, the descriptors of language proficiency most often adopted are those developed by the American Council of Teachers of Foreign Languages (ACTFL). Figure 6.1 shows what the ACTFL descriptors suggest that a student should be able to accomplish in writing at the advanced level.

ACTFL Descriptors for *Writing*

Advanced

Able to write routine social correspondence and join sentences in simple discourse of at least several paragraphs in length on familiar topics. Can write simple social correspondence, take notes, write cohesive summaries and resumes, as well as narratives and descriptions of a factual nature. Has sufficient writing vocabulary to express self simply with some circumlocution. May still make errors in punctuation, spelling, or the formation of nonalphabetic symbols. Good control of the morphology and the most frequently used syntactic structures, e.g., common word order patterns, coordination, subordination, but makes frequent errors in producing complex sentences. Uses a limited number of cohesive devices, such as pronouns, accurately. Writing may resemble literal translations from the native language, but a sense of organization (rhetorical structure) is emerging. Writing is understandable to natives not used to the writing of non-natives.

Figure 6.1 ACTFL Descriptors for *Writing*

The ACTFL guidelines suggest that students should be able to write *correspondence, summaries, narratives* and *descriptions* and engage in *note-taking*. In terms of language features, we are told that the learners need to *join sentences, control morphology and frequent syntactic structures such as coordination and subordination, and use a limited number of cohesive devices such as pronouns*. This is a typical way that standards documents currently define and describe differences in proficiency levels, using different *tasks* or *genres* to identify situations of language use, and then making general statements about a level of complexity or fluency expected, without much information about the particular aspects of language that might be fruitfully focused on at that level.

In California K-12 school settings, English Language Development (ELD) Standards specify what English learners should be able to accomplish at different levels as they gain English proficiency. Elements of the advanced level standards are presented in Figure 6.2.

Like the ACTFL descriptors, the ELD descriptors identify target text types, here *persuasive and expository compositions*. Components of persuasive and expository texts, including *a clear thesis, organized points of support and counter-arguments*, are also specified. From the point of view of the language needed for these tasks, we are told that students should be able to revise their writing for *word choice and organization, consistent point of view and transitions*, and to write *coherent paragraphs through effective transitions and parallel constructions*.

We can see that these proficiency descriptors give information about the tasks students are to do and attempt to specify the level of accuracy or fluency expected. Both sets of descriptors identify situations, functions and tasks that index advanced language use. They also focus on accuracy, with reference to errors, nativeness and the *standard* language. But the information teachers and assessors are given about the language features or grammatical structures that enable students to accomplish the tasks or functions is vague and general. *Complex sentence constructions, cohesive devices, transitions* and *parallel constructions* are named, but little else is specified related to the linguistic features that construct these advanced texts and tasks. This leaves teachers and assessors with little information to guide curricular decisions about appropriate foci for language instruction. As current research increasingly calls for a focus on form (e.g., Doughty and Williams 1998; see Byrnes 2007),

California ELD Standards for *Writing* – Advanced

Advanced

Write persuasive and expository compositions that include a clear thesis, describe organized points of support, and address counter-arguments. Produce writing that establishes a controlling impression or thesis. Structure ideas and arguments within a given context giving supporting and relevant examples. Revise writing for appropriate word choice and organization, consistent point of view, and transitions, which approximate standard grammatical forms and spelling. Create coherent paragraphs through effective transitions and parallel constructions. Edit writing for conventions of writing to approximate standard grammatical forms.

Figure 6.2 California ELD Standards for *Writing* – Advanced

identifying the features most relevant for such a focus is an important issue. In addition, if teachers are to do more than correct the errors students make, they need guidance about how to proactively scaffold language development by helping students adopt new ways of writing.

The descriptors suggest that specifying the focus of language development at different proficiency levels depends on taking the contexts of use as a starting point. These contexts are the most clearly defined aspects of advanced language proficiency, referring to the tasks, genres, assignments and language situations that establish the expectations for language use by students at advanced levels. But, to enhance the specification of tasks with a more complete description of the language features that enable students to accomplish the tasks, we need to link the contexts and tasks with their grammatical realizations.

Systemic functional linguistics (SFL) is a theory of language that offers tools for identifying the linguistic features that are relevant in the construction of different kinds of texts.[1] Different choices from the grammar accomplish different kinds of things for speakers and writers, and the theory enables us to associate linguistic choices with the contributions they make to three kinds of meanings; *ideational* meanings that build the *field*, or content, of a text; *interpersonal* meanings that construe the *tenor*, attitudes, role relationships and evaluation in a text; and *textual* meanings that construct the *mode*, or flow of information, in a text (Halliday and Matthiessen 2004). SFL has been used to identify how language learners draw on linguistic features in constructing different text types (e.g., Er 1993; Hood 2004; Jones *et al.* 1989; Martin 1996; Schleppegrell 1998, 2002). This chapter demonstrates how analysis of texts written in a context of advanced language use can identify the linguistic features that are functional for realizing that context and specify for teachers and assessors the particular linguistic foci that can be associated with advanced proficiency.

Student writing in history

The texts used here to illustrate some linguistic features of an advanced language task are expository essays written in an eleventh grade California history class of English learners.[2] The essays, first drafts written under time pressure, illustrate the resources that these students bring to this task and suggest resources that they still need to develop. The students were responding to a prompt assigned after California's recall election in the fall of 2003 that asked: *What is your view of the recall election? Will it be good for the economy? Will it strengthen democracy in California?* Two essays are analysed here to illustrate the functional contributions of particular language resources to this task and to suggest how teachers can respond to such writing in ways that scaffold the further development of this written genre. Discussion highlights how the linguistic choices the writers make contribute to the presentation of content, to the projection of a stance and to the construction of a well-organized text.

The student writers vary in their control of the resources of English and in their ways of responding to the prompt. While the texts clearly have many of the infelicities ('errors') characteristic of second language writing, these features are not in focus here. Error is a natural part of second language development, and a focus on error can be counterproductive in drawing attention to formal features that may not be crucial to meaning-making while at the same time ignoring language that may be formally correct but ineffective in constructing an authoritative or well-organized text. A primary focus on error can also discourage students from attempting more complex writing patterns. So the focus here is on identifying the strengths writers bring to the writing task and the additional linguistic resources that they could develop to write more effectively.

Of course, particular texts illustrate only certain aspects of the range of language resources that might be drawn on to construct an argument essay, and other examples would bring other resources into focus. But the texts presented here are indicative of the types and range of resources used by students in a larger corpus of 345 texts from which these examples are drawn (Schleppegrell 2005), and the language features in focus are those that are functional for accomplishing the purposes and goals of this writing task.

The resources in focus

The general context of expository writing requires that the writer *displays knowledge authoritatively*, structured in a *well-organized* text (Schleppegrell 2004a). Some key linguistic resources for construing these ideational, interpersonal and textual meanings are presented in Table 6.1:

Table 6.1 Linguistic resources for exposition

- Nominal expressions for naming the points to be made
- Verbs that construct relational processes for defining key terms
- Modality for constructing possibility and necessity in making judgements
- Markers of consequential relationships (*purpose, condition, cause, concession*) for drawing conclusions or supporting assertions
- Projection through verbs of thinking and saying for citing others or taking a stance
- Thematic choices that enable smooth progression in presenting information
- Internal connectors for signposting the organization of the text

Students need to draw on nominal structures that name the arguments to be developed in their texts, both for purposes of delineating the content to be presented and in order to provide an organizational scaffold. Organizing vocabulary (nouns that can serve as superordinate terms) enables the writer to identify the points that the essay will then take up and develop, often using what Flowerdew (2003) calls 'signalling nouns' such as *reason, principles, features,* and other terms that abstract and condense an idea that is going to be developed in the essay. Further, in developing the points to be made, it is often useful to define key terms. While various linguistic resources can be drawn on for this purpose, a repertoire of verbs that construe relational processes (processes of *being* or *having* in attributive or identifying clauses) enables students to write about, for example, what something *means, indicates, includes, involves* or *is associated with.*

Since the essay prompt calls for the students to present an opinion, the writers need to use grammatical options that construct evaluation and judgement in an authoritative way. Resources that serve this purpose include modality that constructs possibility and necessity and markers of consequential meanings that help construct the explicit point of view and argument of the writer. In addition, projection through mental and verbal processes (verbs of *thinking* and *saying*) enables the writer to cite authorities that support or challenge the argument and to present the writer's own stance toward the question (e.g., *Some argue . . .*; or *I believe . . .*).

A construct of functional grammar that enables us to assess how information is presented and built up in a text is *theme/rheme* progression (see also the extensive discussion in Ryshina-Pankova, this volume). Thematic choices help a writer structure information so that key points are highlighted in an expository essay (Martin 1996). *Theme* is identified as the first ideational element in the clause (Halliday and Matthiessen 2004), and it typically introduces information that is presented as known or *given*, with *new* information highlighted at the end of the clause, in the *rheme.* By exploiting the potential to introduce new information in a clause rheme that is taken up again in a subsequent clause theme, the writer can construct a flow of information by distilling what has been presented in the rheme into a nominal or clausal element that can then serve as the point of departure for further discussion and development.

Text organization is also signalled by *internal connectors* that signpost the unfolding of the argument and the structuring of the text. Internal connectors include conjunctive links such as *first, finally,* as well as cohesive demonstratives and other pronouns that refer back to points that have already been made so that the writer can draw conclusions about them, using, for example, *that means; this shows.*

The students' texts presented below illustrate responses to the essay prompt that draw on these linguistic resources in different ways and thereby construct different types of texts. The first example uses nominal expressions and definitions to construct a clear structure and present focused content; but a lack of modality and consequential relationships in the text indicates the weakly developed stance of the writer. Text Two takes a clear position on the question, using strong modality, markers of consequential relationships and rheme-to-theme progression. However, the writer does not use language resources that would organize the text rhetorically, including nominal expressions that name the points to be made, definitions and internal connectors; as a result, the argument does not present the authoritative voice that is most highly valued in such writing. The next sections examine the writers' linguistic choices in more detail.

Text One: clear structure, weak stance[3]

Text One

I will write about the recall, about the Governors, and about is it right or not to recall.

Recall is that the citizens didn't like the Governor and they want to have another governor. And in this time it happened that the citizen didn't like the Grey Davis and they believe that the economy will be better if they will recall the Governor and they said that the Davis for the state's economic and energy problem so that's why they recall him.

The Governors is a people who represents the state who is the person who governs economy and all other things. This time we had a Grey Davis and they recall him and now is a Arnold Schwarzenegger is a Governor.

I think it is not very good to recall the governor for California because it in first time that they recall the Governor it might be a problem for the California, and California's people don't like it very good.

That is the recall the Governor and the thing that they did is not very good for California.

The writer of Text One has attempted to structure the essay in ways that are valued in exposition, beginning with an introduction that names the points to be developed (*the recall, the governor,* and *about is it right or not to recall*). The essay proceeds to develop these three points paragraph by paragraph, and ends with a summary sentence that once again identifies the three issues that the writer has dealt with. So, while infelicitous in some ways, we can see that the text has a structure that shows understanding of academic expectations for explanation and persuasive argument.

In naming the points to be developed, however, the student does not have the nominal resources to *name* the third point, and instead uses the clausal structure *is it right or not to recall.* The ability to construct such meanings in nominal expressions, rather than in whole clauses, is an important feature of advanced language development. Using abstraction and *grammatical metaphor* (Halliday 1993), the discussion of *is it right or not to recall* could be recast, for example, as *my opinion about the benefits of and problems with the recall,* contributing simultaneously to a more authoritative stance and to a more felicitous text.

While the text has a clear structure, it lacks a clearly presented thesis that gives a purpose for the writing. An important move in the introduction to an essay, a thesis statement typically draws on modality and consequential markers to propose and support a position (Schleppegrell 2004a: 101–02). As is shown below, this writer uses few of these linguistic resources, whose development is crucial for expository writing.

Within each of the first two body paragraphs, the writer defines *recall* and *governor,* highlighting the need for effective strategies for definition. He then uses a strategy of temporal organization to present background about the recall and the governor in each paragraph, using temporal and additive conjunctions to structure the paragraphs as a *story* about the recall election (*And in this time it happened, and they believe, and they said, This time,* and *and now*). These definitions and resources for temporal organization are useful for developing background, but the student needs to do much more than define and recount what has happened.

The writer needs to make judgements about the pros and cons of the recall election, and to do so needs to draw on markers of consequence. We see one such attempt in the *so that's why* that provides a reason for the recall in the second paragraph. The second paragraph also presents some evidence for this conclusion by telling what citizens *didn't like, believed* and *said,* demonstrating the value of these mental processes for bringing the voices of others into the text. Projection through mental and verbal processes such as *believe, think, know, discuss, analyse,* enables the writer to use what others have said as evidence. The processes constructed in these verbs also enable the writer to present his own view (what he *will write* and what he *thinks*). But the essay lacks the consequential markers and structuring elements that enable a claim to be presented and supported. So while the text states an opinion, it does not construct an argument with claims and evidence, using the linguistic resources that would enable such construction.

The student does state a position with support in the fourth paragraph, using projection through a mental process (*I think*) and the consequential marker *because,* but the modality of possibility (*it might be a problem*) attenuates the judgement. As we see below, control of modality is an area of meaning that is quite challenging for language learners, as it constructs judgements about possibility and necessity that are crucial for sounding reasoned and authoritative.

The last paragraph, a sentence that begins with the cohesive demonstrative *that,* illustrates again the student's strong sense of rhetorical organization. The cohesive *that* is an internal connector that refers back to the whole text as a point of departure for this concluding sentence. The student is clearly aware of the rhetorical expectations for a text of this type, and can continue to build on that awareness as his writing develops by incorporating the interpersonal stance and judgements that call on a broader range of linguistic resources.

Text Two: hortatory stance, emergent structure

Text Two

The recall must be good for California, since more than half of Californians voted for the recall of Governor Davis. California's budget was in crisis while he was in the office, and something was need to be done.

Since something was need to be done, people voted 'yes' on the recall. When people voted 'yes' on the recall, I think they knew that they were doing, and since Governor Davis was recalled, that means that many people were not satisfied with the way he governed their state.

Many taxes did not fix the budget, but made even more people to vote 'yes' on the recall. And since there were no improvements in the California's budget, Davis was removed from the office. There are other ways to improve the budget, then taxing.

Arnold Schwarzenegger was elected the new governor because people believed that he can fix the budget and bring California back to prosperity, and if he would find some ways to do that without making more damage, then the recall will be definitely good for California.

Since many businesses left the state because of heavy taxes, with the new governor who promised to reduce some taxes, the businesses that left might come back, and more businesses would be opened without fear of losing profits because of high taxes. This would much improve the budget and help California economically.

As I already mentioned, I think the recall will be good for California, politically and economically.

The writer of Text Two has adopted a different strategy: presenting many claims with support for them, but structuring the essay in a very emergent way, with no hierarchy foreshadowed by nominal elements that name points. Instead, this essay moves from clause to clause in what we might characterize as an oral style. Assertions are presented with strong subjective modality of necessity (what *must be* and that something *was need to be done*) (Halliday and Matthiessen 2004, Schleppegrell 2004b) and consequential relationships constructed with *since, that means, because*, and an *if-then* construction support the assertions. This pattern of modal judgement followed by clauses introduced by causal connectors constructs a hortatory style of the type Crowhurst (1990) identifies with less mature writing. Text Two uses projection to bring in what people *knew* and *believed* and projects the writer's own view using *I think* in the final paragraph, but the writer does not define anything.

The kind of structuring this writer draws on is a *rheme-to-theme* presentation that builds the argument he is making. At the beginning of the second paragraph, the theme *since something was need to be done* repeats the new information from the rheme of the previous clause so that the consequence of this can be drawn. Similarly, the consequence presented in the next clause, that people voted 'yes' on the recall, is re-presented in the subsequent theme so that the writer can make a judgement about it. In each case, however, it is the exact wording that is repeated. The writer uses a similar strategy in the third paragraph, but here the theme of the first sentence, that *many taxes did not fix the budget*, is picked up and re-presented in the next sentence as *there were no improvements in the California's budget*, using the nominal *improvements* to condense the clausal *did not fix*, a more sophisticated use of the rheme-to-theme progression that uses the nominal construal. The writer of Text Two also uses the internal demonstratives *that* and *this* to make links and draw conclusions.

The writer draws on the modality of possibility to move from justifying the recall election to predicting what might happen as a result in paragraphs four and five. The *if-then* construction sets up the conditions under which the recall *will definitely be good*, and the next paragraph highlights what *might* and *would* happen under the new post-recall scenario that the writer lays out. We see how the modality of necessity and possibility enables the writer to say what *must be done* and to present potential outcomes. However, getting control of expression of degrees of necessity and possibility requires learning a range of modal meanings and their realizations (see Lock 1996). Alternative meanings of the modal *must*, for example, create the infelicity represented in the writer's *The recall must be good for California*, where the epistemic, less authoritative meaning is evoked. The assertion is presented as a deduction, with the voters' approval of the recall given as evidence for the writer's belief (It must be good because they voted for it). While this might be read as flawed reasoning, the learner's inexperience with the range of meanings construed by the modal must also be taken into account. Students' logical reasoning is constructed in language, after all, and the meanings at stake here are clearly nuanced and advanced. This writer marshals evidence and arguments for his point of view, but the high modality of necessity and the structuring through alternating assertion and support, using mainly the causal conjunction *since*, constructs a tone that does not enable authoritative presentation of the arguments. Using organizing vocabulary to name the claims to be made in abstractions such as *budget issues* or *excessive taxes*, a more effective use of modality, and control of a wider range of resources for consequential meaning would enable the writer to argue in a way that would be more highly valued in the contexts of advanced literacy tasks such as this.

Implications

The writer of Text One uses organizing strategies that serve well to construct exposition, but does not develop claims supported by evidence, and so ultimately fails to produce an effective argument. The lack of modality and consequential connectors in Text One indicate that the text does not make the judgement that is called for. Text One uses causal connectors only to support the one statement of the writer's point of view and to say why voters recalled the governor. The writer of Text Two, on the other hand, draws on resources for modality and consequential relations and, in doing so, constructs a set of claims supported by evidence. However, the modality and strategy for rhetorical organization he employs result in a hortatory response that may not be highly valued in an academic context.

Both writers use patterns that can make functional contributions to the development of an argument, but they need assistance in developing a better sense of when and how these patterns can most effectively be deployed. For example, the nominal elements that name the arguments to be developed help scaffold the organization of a text, and the ability to construct a nominal element that summarizes or recaps a point that has been developed also enables the student to present judgements using rheme-to-theme progression. In text-organizational terms, the construction of a complex nominal element or use of internal connectors allows information that has been presented in a series of clauses to be distilled and further commented on, enabling a chain of reasoning to be developed by the writer, a strategy that Christie (2002) links with contexts of advanced language use. This suggests that development of relevant organizing vocabulary and a focus on how nominal elements can facilitate text structuring would be a fruitful focus for pedagogy at advanced levels.

Helping students see where definitions are appropriate and modelling various strategies for their construction is also relevant for this kind of writing. A focus on developing control of nuanced use of modality and a repertoire of resources for marking consequential relationships can assist students both in the construction of a thesis and in making claims and providing supporting evidence. Projection of the thinking and saying of oneself and others can contribute evidence and perspectives that help make an argument, but writers also need to learn ways of introducing such perspectives that go beyond heavy reliance on simply saying what *I* and *others think* and *say*. The temporal organization that the writer of Text One has adopted is useful for presenting background information, but a different organizational strategy is needed to present claims and evidence, and internal connectors and thematic progression can be a focus of pedagogy to help students see how these linguistic resources help organize a text.

These are resources and strategies that can be brought to students' conscious attention, enabling teachers to be proactive in helping students develop the linguistic resources they need for exposition. Students can learn to develop a thesis and consider whether they need to define and expand the terms they introduce, focusing on the linguistic resources that enable these moves. They can analyse the claims and evidence they present and consider how their thematic choices are contributing to the overall structuring of information in the texts they write. Neither of these writers produced the counter-arguments called for in the ELD Standards for advanced writing, suggesting that teachers could also profitably introduce students to linguistic resources that enable concession and refutation at this level.

The *organization, point of view* and *coherence* called for in advanced writing in the ELD Standards are realized in the linguistic choices writers make. This points to the need for teachers to ground their writing instruction not only in a discussion of the *purposes, audience* and *organization* of a text, but also in a focus on the language features that enable the purpose to be realized, the audience expectations to be met and the organization to be constructed. By focusing on the linguistic features that are functional for particular tasks, teachers provide students with the tools they need to develop advanced language proficiency.

Conclusions

Advanced learners often have already developed literacy in another language, and so may be aware of the potential in language for construing meanings in abstract and academic ways. Now, as second language learners, they are challenged to reinterpret this experience in the new language. By focusing on the meaning potential that learners are exhibiting and offering language input to expand that meaning potential, we recognize the challenges of advanced literacy by seeing the overall patterns of language that students need to develop to effectively accomplish such advanced tasks as constructing expository writing that makes claims, presents evidence and makes judgements in authoritative ways.

The ACTFL guidelines and ELD Standards indicate what is expected of learners in terms of genre and tasks, but are largely silent on linguistic features that might enable writers to accomplish those tasks. What the teacher can do to build the language, or what the assessor can look for in terms of development of language itself is left vague and underspecified. This chapter has suggested that we can expand descriptions of advanced language to include more specific information about the linguistic resources that teachers and learners could focus on, and that language assessment might target, as students gain in proficiency. Descriptions of advanced language proficiency specify contexts of use that a functional linguistic theory allows us to link with the grammatical resources needed to construe the contexts in effective ways. Recognizing the functionality of the grammar for making meaning, we can identify the language resources that teachers and students can focus on as they work on abstract and complex texts and tasks.

Learning a new language is a way of expanding one's meaning potential to new contexts, so a focus on contexts of meaning is crucial for developing language to advanced levels. By identifying the linguistic resources that are functional for meeting the expectations of particular tasks, learners' movement into more effective use of those resources can be charted and scaffolded. Such an approach enables us to focus on language development related to the contexts in which students will use the language they are learning. Byrnes (2002: 426) calls for a greater orientation to language meaning and use, pointing out that 'If programs are to . . . recognize the complexly staged, long-term process of successive approximative interlanguage systems that learners follow, they need ways of envisioning what counts as "success", both from the teachers' and from the learners' perspective, without relying on the deceptive certainty that goes with accuracy.' A functional linguistics approach that recognizes the meaning-making potential of different language choices focuses us on the meanings that learners are constructing, and not simply on the errors that they will inevitably continue to make as they expand their meaning-making into new contexts.

Acknowledgement

This research was completed with support from the University of California Linguistic Minority Research Institute (UCLMRI) and with the cooperation of the History Project at UC Davis and teachers from the Grant Joint Union High and Sacramento City Unified School Districts. I gratefully acknowledge their contributions.

Notes

1 For accessible introductions to functional grammar see Butt *et al.* (2000); Eggins (2004); Droga and Humphrey (2002); Thompson (2004).

2 These texts were gathered by the Area 3 History and Cultures Project, a professional development project that is part of the California History/Social Sciences Project (see http://historyproject.ucdavis.edu/). The essays come from a corpus of 345 essays written by 8th and 11th grade students that were analysed as part of a larger study (Schleppegrell 2005).

3 The texts are reproduced as the students wrote them, but with the spelling corrected for ease of reading.

References

Butt, D., Fahey, R., Feez, S., Spinks, S. and Yallop, C. (2000) *Using Functional Grammar: An Explorer's Guide* (2nd edn) Sydney: National Centre for English Language Teaching and Research, Macquarie University.

Byrnes, H. (2002) 'The role of task and task-based assessment in a content-oriented collegiate foreign language curriculum'. *Language Testing*, 19, 419–37.

Byrnes, H. (in press) 'Language acquisition and language learning', in D. Nicholls (ed.), *Introduction to Scholarship in Modern Languages and Literatures* (3rd edn). New York: MLA.

Christie, F. (2002) 'The development of abstraction in adolescence in subject English', in M. Schleppegrell and M. C. Colombi (eds), pp. 45–66.

Crowhurst, M. (1990) 'The development of persuasive/argumentative writing', in R. Beach and S. Hynds (eds), *Developing Discourse Practices in Adolescence and Adulthood.* Norwood, NJ: Ablex, pp. 200–23.

Doughty, C. and Williams, J. (eds) (1998) *Focus on Form in Classroom Second Language Acquisition.* Cambridge: Cambridge University Press.

Droga, L. and Humphrey, S. (2002) *Getting Started with Functional Grammar.* Berry, NSW, Australia: Target Texts.

Eggins, S. (2004) *An Introduction to Systemic Functional Linguistics* (2nd edn). London: Pinter.

Er, E. (1993) 'Text analysis and diagnostic assessment'. *Prospect*, 8 (3), 63–77.

Flowerdew, J. (2003) 'Signalling nouns in discourse'. *English for Specific Purposes*, 22, 329–46.

Halliday, M. A. K. (1993) 'Towards a language-based theory of learning'. *Linguistics and Education*, 5, 93–116.

Halliday, M. A. K. and Matthiessen, C. M. I. M. (2004) *An Introduction to Functional Grammar* (3rd edn). London: Arnold.

Hood, S. (2004) 'Managing attitude in undergraduate academic writing: a focus on the introductions to research reports', in L. J. Ravelli and R. A. Ellis (eds), pp. 24–44.

Jones, J., Gollin, S., Drury, H. and Economou, D. (1989) 'Systemic-functional linguistics and its application to the TESOL curriculum', in R. Hasan and J. R. Martin (eds), *Language Development: Learning Language, Learning Culture.* Norwood, NJ: Ablex, pp. 257–328.

Lock, G. (1996) *Functional English Grammar: An Introduction for Second Language Teachers.* Cambridge: Cambridge University Press.

Martin, J. R. (1996) 'Waves of abstraction: organizing exposition'. *The Journal of TESOL France,* 3, 87–104.

Ravelli, L. J. and Ellis, R. A. (eds) (2004) *Analysing Academic Writing: Contextualized Frameworks.* London: Continuum.

Schleppegrell, M. J. (1998) 'Grammar as resource: writing a description'. *Research in the Teaching of English,* 32, 182–211.

Schleppegrell, M. J. (2002) 'Challenges of the science register for ESL students: errors and meaning-making', in M. J. Schleppegrell and M. C. Colombi (eds), pp. 119–42.

Schleppegrell, M. J. (2004a) *The Language of Schooling: A Functional Linguistics Perspective.* Mahwah, NJ: Lawrence Erlbaum.

Schleppegrell, M. J. (2004b) 'Technical writing in a second language: the role of grammatical metaphor', in L. J. Ravelli and R. A. Ellis (eds), pp. 172–89.

Schleppegrell, M. J. (2005) 'Helping content area teachers work with academic language: Promoting English Language Learners' literacy in history'. Santa Barbara, CA: UC Linguistic Minority Research Institute.

Schleppegrell, M. J. and Colombi, M. C. (eds) (2002) *Developing Advanced Literacy in First and Second Languages: Meaning with Power.* Mahwah, NJ: Lawrence Erlbaum.

Thompson, G. (2004) *Introducing Functional Grammar* (2nd edn). London: Arnold.

7 Grammatical metaphor: academic language development in Latino students in Spanish

M. Cecilia Colombi

This article describes a particular lexicogrammatical resource that Spanish uses to realize academic language, the resource that Systemic Functional Linguistics (SFL) refers to as grammatical metaphor (GM). Developed mainly by Halliday (1994), the notion of GM represents an original and innovative contribution that identifies and describes the fact that scientific and academic registers, in writing and in speaking, are functionally oriented to accomplishing 'objectification' and 'abstraction' of their content. They achieve this functional goal through the linguistic means of GM, a resource that condenses information by expressing experiences and events in an incongruent form, as contrasted with the more customary congruent form that prevails in everyday language use. The paper presents three types of GMs as a way of explicating and tracing the development of academic language at the college level in heritage speakers of Spanish: (1) *Ideational GM*, an incongruent representation of experiential meaning; (2) *logical GM*, a way of organizing ideas at the level of discourse in an incongruent manner; and (3) *interpersonal GM*, which presents authorship in the text both implicitly and explicitly.

I have chosen the Spanish heritage learner for several reasons. First, as a consequence of recent demographic trends,[1] Spanish language use has visibly and audibly increased in the United States. As new immigrants interact in the community, in schools, businesses and the workplace, in Spanish with those who have been here for some time or who were born here, Spanish is not only heard with greater frequency in public environments, it is also seen more prominently in the media and in advertising (Carreira 2003). This demographic change and the increasing use of Spanish in public and private settings has important implications for language teachers and students. My home institution, for example, has seen a considerable rise in the number of Latino students who are pursuing Spanish for professional purposes.[2]

Second, although some students are able to develop the desired public and academic registers on their own, primarily through writing and reading, a more adequate theoretical and pedagogical framework than that currently informing academic language development is needed if a greater number of students is to be successful. In other words, language educators need a way of understanding and teaching how language means in academic contexts. Christie (2002a), Ravelli and Ellis (2004), Schleppegrell (2004) and Unsworth (2000), among others, have emphasized the need to focus explicitly on how language means in academic contexts and have pointed to SFL as a viable educational linguistic framework to address issues of genre and register in the classroom. This is so because there is ample evidence that students, in general, will develop academic-level proficiency primarily through language-based interactions in school settings; this insight is particularly true for Spanish language arts instruction in the context of minority language teaching in the United States.

From the linguistic point of view, studies have shown us that the expansion of bilingual competence in a heritage language, especially at the academic register, helps the development of English as an academic language (Cenoz and Genesee 1998; Cummins 2000; Harley *et al.* 1990; Valdés 2001). Cummins (2000) has studied the bilingual proficiency of heritage students in schools for an extended period of time. He was the first to suggest two categories for bilingual competence: a conversational language capacity (BICS: bilingual interpersonal communicative skills) and a cognitive/academic language capacity (CALP: cognitive academic language proficiency). He suggested that there was a common underlying language proficiency and that linguistic skills could be transferred from one language to the other. As a consequence, he claimed that it is **easier** and **faster** for heritage speakers to develop academic/cognitive skills in their heritage language first and then transfer those skills to the second language. In other words, academic language skills developed in the first language facilitate their development in the second language, especially at advanced literacy levels (Belcher and Connor 2001; Beykont 2002; Schleppegrell and Colombi 2002; Valdés 2001, 2003). Studies have also shown that bilingualism and biliteracy increase the cognitive abilities of students (August and Hakuta 1997, 1998; Cummins 2000).

Although many researchers have emphasized the need to develop the academic/professional register of heritage language speakers to enable them to learn English faster and more easily, few have focused on the linguistic development of Spanish as a heritage language. Of those, the majority have dealt with written language (Acevedo 2003; Colombi 1997, 2000, 2002, 2003; Gibbons 1999; Martínez 2003; Schleppegrell and Colombi 1997); and only a few have analysed oral language (Achugar 2003; Valdés and Geoffrion-Vinci 1998).

In this paper I endeavour to address the concept of GM as a distinctive linguistic characteristic of Spanish academic texts, oral and written. My interest in doing so is this: texts with a high degree of GM tend to be considered prestigious in U.S. culture, and the use of GM is considered an essential marker of academic and professional-level literacy. As Spanish becomes much more present in the public sphere in the United States an explicit pedagogy for Spanish as an academic language becomes a critical aspect of equity, access and literacy in the public square. To the extent that the analysis presented here supports effective explanations of Spanish academic texts as they are used in school and, furthermore, to the extent that it is possible to specify pedagogies that support the acquisition of key features of such language use, these insights could contribute to setting an agenda for the curriculum in Spanish as a heritage language in the United States, thereby serving different groups of students who are engaged in the acquisition of advanced literacy for a variety of purposes.

Linguistic features of academic language

In a longitudinal study of Spanish as a heritage language in the United States, I followed students' writing and oral presentations for a period of a year (three academic quarters) in a program of Spanish for Native Speakers (SNS). This program follows an eclectic pedagogical approach, combining a text-based curriculum with a Freirean (problem-posing, peer-tutoring and identity-related activities) with a process-oriented methodology (e.g., multiple version assignments, peer-editing and journals). The first versions of all students' compositions and their oral presentations were collected, transcribed and analysed following an SFL framework.[3] A look at the development of these Latino students' writing in Spanish in an academic context (Colombi 2000, 2002, 2003) shows a progression along a continuum of expressive forms, from what one might, quite generally, refer to as a colloquial register to more academic forms of language use. The characteristics listed in Table 7.1 are useful for defining language use along that continuum.

Because SFL builds on the fundamental interconnectedness of language use (including specific forms of language use) and the social context, a key aspect of an educational approach that uses insights from SFL is to assure that students develop exactly this: an awareness of the fundamentally social nature of language use practices – and that includes literate practices – along with an awareness of how these practices are socially positioned. For that reason, a commonly postulated difference between oral and written modes of language must always be examined in context, in order that one may

Table 7.1 The oral–written continuum (adapted from Halliday 1985)

Oral ⟵⟶	***Written***
Linguistic characteristics	
➢ dynamic structure	synoptic structure
➢ everyday lexicon	specialized lexicon
➢ non-standard grammar	standard grammar
➢ grammatical complexity	high lexical density

understand the local considerations that motivated specific language choices (see Halliday 1985 and Chafe and Danielewics 1987, among others, for such studies).

Halliday (1998) points to lexical density, nominalization and grammatical metaphor as the main lexicogrammatical characteristics of written (academic) language. Indeed, while the idea of lexical metaphor in a conventional sense is generally available in SFL theory, it is the notion of GM, developed mainly by Halliday (1994), which represents a particularly original and innovative contribution to linguistic theory. Specifically, Halliday (1993) proposes that 'young children's world of meaning is organized congruently', i.e., their language reflects directly their experience of the world. However, as they approach adolescence and adult knowledge, young people begin to reconstrue 'their clausal grammar in a different, nominalized form', what he calls *grammatical metaphor*, a process that is strongly influenced by schooling. Directly related to experiential knowledge, GM nevertheless indicates a shift from commonsense ways of meaning-making, where the lexicogrammatical forms chosen are congruent with the semantics of the event or experience, to uncommon ways of meaning-making through a more metaphorical reconstrual of experience.

Moving into the educational realm, Christie (2002b: 46) explains that it is in secondary instruction that adolescents start handling 'the building of generalizations, abstraction, argument, and reflection on experience that advanced literacy seems to require'. In particular, she suggests that children come to school with an understanding of the kind of grammatical generalization that allows them to interpret and handle common sense and interpersonal language, with grammatical abstraction evolving through schooling in the primary years. However, it is only in schooling at the secondary level that young adults start developing what would count as advanced literacy through the use of grammatical metaphor. Table 7.2 lays out that development.

Inasmuch as GM is a linguistic resource that condenses information that is otherwise expressed in congruent ways, the use of metaphorical forms represents a choice. It signals the value the discourse communities that are engaged in such language use attribute to 'objectification' and 'abstraction', and how they achieve that functional orientation through the use of GM that packs more information into a clause. This choice is particularly prevalent in scientific or academic registers, where the informational density achieved through GM has been particularly well studied for English in the sciences, in

Table 7.2 Stages of language development (adapted from Halliday 1993 and Christie 2002b)

• Grammatical generalization	⟶	Interpersonal language (common sense)
• Grammatical abstraction	⟶	Basic literacy
• Grammatical metaphor	⟶	Advanced literacy

history and language arts (Eggins *et al.* 1993; Halliday 1998; Martin 1993, 1996; Simon-Vandenbergen *et al.* 2003). By contrast, little comparable work exists in Spanish (but see Gibbons 1999; Colombi 2000, 2002).

Grammatical metaphor as a linguistic resource in academic language use

Accordingly, this paper investigates the use of GM in Spanish in order to begin to address this lacuna in the particular context of academic Spanish. To repeat, the three major types of GM, the ideational, logical and interpersonal GM, occur when the usual or 'congruent' realization of meaning is given a 'non-congruent' or metaphorical expression: ideational GM relates to experiential meaning, logical GM construes textual meanings and interpersonal GM creates interpersonal meanings. In the following, I will explore these three forms at some depth for their various meaning-making potentialities and in terms of their various formal manifestations.

Ideational grammatical metaphor

Among the ways of representing experience SFL highlights the following:

Example 1: Congruent realization of meaning

En esta novela	Poniatowska	cuenta	la historia de una mujer mexicana . . .
↓	↓	↓	↓
Circumstance	Participant	Process (verbal)	Thing (Ana 33–01)[4]

By contrast, the sentence below shows a more incongruent form of representing reality through a GM.

Example 2: Incongruent realization of meaning: Grammatical metaphor

La liberación femenina	no fue	un tema que afrontó el pueblo mexicano durante la Revolución.
↓	↓	↓
Carrier	Process (relational)	Attribute (Ana 33–01)

Why consider this a metaphorical expression? The answer is predicated on accepting the notion that the congruent form is the unmarked way in which we represent experience and that the alternative or marked realization is a form of metaphor. Thus, in this example a process that would normally be expressed through a verb (*liberar*) has been metaphorically transformed into a fixed object, expressed by a noun (*liberación*). Table 7.3 is an adaptation of Halliday's detailed description of GM in English (1998: 211).

Table 7.4 presents the *metaphorical* change in the grammatical class category and the semantic type. Semantically, the resulting grammatical category (e.g., *liberación*) maintains the characteristics of the noun and of the process. In the ideational GM in Spanish as in English, movement is from left to right, i.e., from the clause complex to the noun.

Table 7.3 Grammatical metaphor (adapted from Halliday 1998)

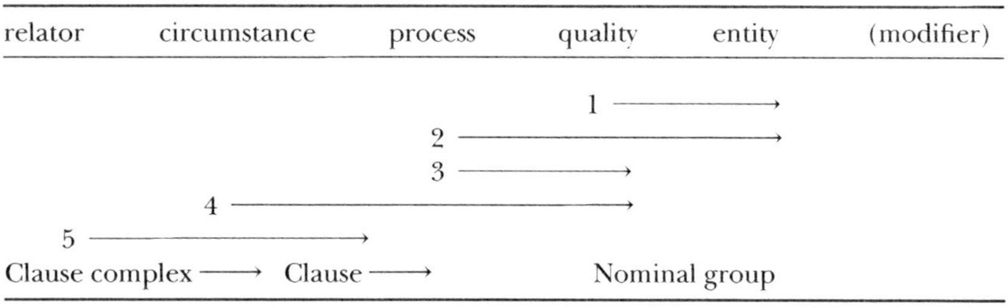

Table 7.4 Class shift (semantic type)

Congruent	**Metaphorical**
1. Adjective (Qualifier)	Noun (Entity)
2. Verb (Process)	Noun (Entity)
3. Verb (Process)	Adjective (Qualifier)
4. Adverb (Circumstance)	Adjective (Qualifier)
5. Conjunction (Relator)	Prepositional Phrase (Circumstance)

Nominalization as a central grammatical metaphor

According to Halliday (1994: 352), 'nominalizing is the single most powerful resource for creating GM'.

Formal realizations of nominalization

Example 3: Nominalization
La **emigración de la época de la Revolución Mexicana** fue de notable importancia para el campo de la literatura, debido a la gran cantidad de intelectuales que pasaron a los Estados Unidos huyendo de la agitación social de México. (M. Martín-Rodriguez, 2001: 227)

Congruent: Verbs (Process)	Incongruent: Nouns (Entity)
(los intelectuales) **emigraron** durante la época de la Revolución Mexicana	La **emigración** de la época de la Revolución Méxicana . . .

Illustrating the central feature of GM, nominalization as a GM combines the feature of a 'process' (*emigrar*) with that of an 'entity' in a 'semantic junction'. According to Halliday, this semantic junction combines the meaning of the semantic type of congruent form (process: *emigrar*) and that of the metaphorical form (entity: *emigración*) into one language form. Furthermore, as in English, GM in Spanish allows for the condensation of information: once the process 'emigrate' has been nominalized it can be expanded considerably. As a result, like its English counterpart, the Spanish nominal group is the most powerful and also the most frequent resource for making meaning in academic texts. Accordingly, my own analysis of Spanish texts finds nominalizations to constitute 70 per cent of all GMs found in the texts. It is thus in line with the findings of Eggins *et al.* (1993), Ravelli (1988) and Jones (1990), who have demonstrated its frequency in English student writing.

The following introductory paragraph from Ana's writing in the third quarter of instruction (i.e., the most advanced course of the SNS series), is a good example of nominalizations:

Example 4: Nominalization – Ana's introductory paragraph after 9 months of instruction

Las reformas dentro de la Revolución mexicana (Ana 33–01)[5]

'¡Tierra y Libertad!' fueron las famosas palabras que gritó Emiliano Zapata que comenzaron la Revolución Mexicana. Antes de que rebeldes, como Zapata, comenzaran a levantar armas ***la división de las clases sociales*** era visible y las inigualdades no se toleraban más por el pueblo. Sin embargo, hubo ciertos aspectos de la sociedad mexicana que quedaron sin solución a finales de la guerra. Por ejemplo ***la liberación femenina*** no fue un tema que afrontó el pueblo mexicano durante la Revolución. Por falta de atención a este tema, se incrementó ***la división entre los roles de los sexos.*** La guerra resultó en ***el aumento del machismo y el retraso de los derechos de las mujeres.***

Eggins *et al.* (1993) explain the functions of this type of GM in history texts in English. The same functions are found in the Spanish model texts students read and interacted with in the SNS courses and, consequently, in their essays:

Functions of nominalizations

a) A central function of nominalization is to 'remove people', as it were. Numerous functions may be embedded in that 'removal', enabling different interpretations of this particular GM: actors are no longer readily identifiable, they play a minor role, they are irrelevant for the case at hand, the writer is unable or unwilling to identify them or, coming from the other side, the end result of an action is more prominent than the action itself. In the examples I first provide a native speaker's use of this kind of GM, followed by the heritage language student's use.

Proclamar el inglés lengua única de los Estados Unidos es una prueba de miedo y soberbia inútiles. (Fuentes 2001: 254)

. . . se incrementó ***la división entre los roles de los sexos.*** (Ana 33–01)

b) A core characteristic of nominalization as a GM is that it turns actions into things, as is illustrated in the following sentence pair:

> **Hablar más de una lengua** no daña a nadie (Fuentes 2001: 254)
> La guerra resultó en ***el aumento del machismo y el retraso de los derechos de las mujeres.*** (Ana 33–01)

c) Nominalizations also give existence to 'things'; in particular, they create conceptual objects:

> **El temor de los legisladores norteamericanos** que condicionan la 'estadidad' a la renuncia de la lengua es, desde luego, el miedo de que, si Puerto Rico mantiene el derecho al español, Texas, Arizona o Nuevo México reclamen lo mismo. (Fuentes 2001: 252).
> Por ejemplo, ***la liberación femenina*** no fue un tema que afrontó el pueblo mexicano durante la Revolución. (Ana 33–01)

When such nominalizations remove the agents of actions, they create more distance between the event and the participants. Then, once the actions have been nominalized, they can be talked about in more 'material' terms, as having occurred, as being available for modification and, most importantly, for movement in conceptual space as actors in their own right.

In the students' writing development in academic Spanish just that kind of movement from congruent language into more incongruent language can be observed. More importantly, the exact development is illustrated as well in their oral language development, as shown in Ana's language use in her final oral research presentation:[6]

Excerpt from Ana's oral presentation of her final research project on soap operas (telenovelas) (A33–OP–03):

15. este . . . entonces ahora . . . este . . .
16. . . . **comenzamos** a . . . a preguntarnos si hay algo más además de entretenimiento en estas . . . este . . . novelas.
17. Y . . .
18. la otra opción verdad que . . . dije YO
19. XXX que mi . . .
20. mi idea sería que. son,
21. es ***una transmisión de valores sociales,***
22. XXX este . . , mediante comunicación en masas.

Importantly, the emerging ability to use GM, in this case, nominalization, is not replacive; rather, noncongruence in GM is to be imagined as existing along a continuum. Thus, Ana realizes the verb-process *transmitir* as a noun ***transmission***, which can be modified (*de valores sociales*). Later on in her presentation she opts for a more congruent form of this nominalization when she says:

27. Entonces este . . .
28. sí es cierto que son . . . valores sociales que son ***transmitidos*** y no . . .
29. este . . . solamente entretenimiento,
30. este . . . ¿qué son los . . . qué son los mensajes que están, ahmm . . .
31. que les estamos diciendo, a los a los ciuda. tele. videntes . . . de las novelas?

Much later in this passage she formulates her thoughts like this:

161. Ah . . . en conclusión, rápidamente . . . este: . . .
162. las la tele, la, la tele. visión no sola . . .
163. un medio de comunicación neutro por lo tanto ***transmite*** contenidos valóricos,
164. este . . . creencias y modelos de conducta,
165. que la hacen esencialmente un sistema educativo.[7]

This type of nominalization constitutes the most frequent ideational GM in Spanish, not only in the academic texts to which students are exposed, but also in their own writing. At the same time, as students start using GM in their writing, the lexical density of their texts also increases, reflecting a more condensed and incongruent form of semiosis and, therefore, of language use.

Verbal processes expressed through adjectives used as grammatical metaphors

The second most frequent type of GM is the verb-process functioning as an adjective. The metaphoric transformation of a process to a qualifier represents a shift from meaning 'construed as process' to meaning 'construed as qualifier', from 'verb' to 'adjective', and from a typical function in the clause of process to that of epithet/qualifier in the nominal group. In Spanish, when the verb-process functions as a qualifier (past participle), it agrees in number and gender with the noun modified. In other words, in Spanish the GM has been *grammaticalized,* inasmuch as the past participle agrees in gender and number with the thing modified, thereby becoming a structural unity.

> En el caso que aquí nos interesa, las conclusiones de Sánchez son en extremo **productivas** para la concepción de la literatura chicana como literatura de resistencia frente a la cultura y sociedad **dominantes** . . . (Martín-Rodriguez 2001: 232)

Congruent: Verbs (Process)	Incongruent: Adjectives (Qualifier)
(las conclusiones) producen (resultados)	(conclusion**es**) **productivas**
(esta cultura y sociedad) dominan	(cultura y sociedad) **dominantes**

As students progress in their development of Spanish academic language during the year, they make increasingly more use of this GM in their writing. This ideational GM, too, allows for condensation of information in a more incongruent form, i.e., agents are removed and time is not defined. The following student writing samples show students using this kind of GM after the first three months of participating in the writing courses.

Ana's essay (33–01):[8]

> En la constitución de 1917, aunque fue muy **progresista**, no se otorgó el derecho de votar a la mujer hasta el año 1954. Carlos Alvear Acevedo, menciona en su artículo 'La Revolución Mexicana' (1993), algunas de las reformas **progresistas** de la constitución: la prohibición de la esclavitud, la libertad de trabajo, el 'juicio de amparo' (405–406) y muchas otras más. Algunos de los planes **mencionados** por Acevedo son El Plan de San Luis, **escrito** por el candidato presidencial Francisco Madero (394), y El Plan de Ayala (399), **escrito** por el rebelde Emiliano Zapata.

Table 7.5 provides a summary of the development of Spanish adjectivization.

In other words, Spanish grammar allows for the semantic junction to be reflected in the grammatical and semantic form of an adjectival GM, conflating the noun and modifier in a unity of meaning and form.

Thus far I have detailed only two types of ideational GMs that have frequent realizations in Spanish: processes remapped as nouns or as adjectives. Table 7.3 above presented the movement from left to right in the realization of the GM, i.e., from more congruent (transparent) language to more incongruent forms. It is important to recognize that relators (or conjunctions) can be construed as a circumstance (i.e., grammaticalized as a preposition, in a prepositional phrase), that a process can be construed as a quality (grammaticalized as an adjective), and, finally, that a process or a quality can be construed as an entity (grammaticalized as a noun), but not the other way around (Halliday 1998: 211).

Table 7.5 Spanish adjectivization: semantic and grammatical junction

. . . la constitución fue muy **progresista** (singular – feminine)

las reformas **progresistas** (plural – feminine)

los planes **mencionados** (plural – masculine)

(el plan) **escrito** . . . (singular – masculine)

Logical grammatical metaphor

Logical grammatical metaphor refers to the condensation of meaning in an incongruent way at the level of the organization of the discourse. The most congruent form of joining two ideas is with a conjunction. By contrast, when conjunctions are realized through processes and nouns, allowing for two or more clauses to become one, that realization is referred to as a logical GM. Once more, the following examples of logical GM are taken from Ana's writing.

> Las palabras del Arzobispo no **causaban** miedo en la gente pobre salvadoreña, sino agradecimiento de que por fin alguien pensaba en ellos. (Ana–32–03)
>
> La guerra ***resultó*** en el aumento del machismo y el retraso de los derechos de las mujeres. (Ana 33–01)
>
> Este poema ***tiene relevancia*** a la situación actual en los Estados Unidos, ya que muchos latinos piensan regresar a sus países de origen después de haber huido por razones políticas. (Ana 32–03)
>
> ***El valor*** de la obra reside en que la gente no pierde la esperanza de algún día regresar a su 'antigua tierra', que es parte tan importante de su vida. (Ana 32–03)

Lorena's development shows a similar trajectory:

> ***La consecuencia de esto es que*** ahora la economía del país está muy inestable. (Lorena 33–05)
>
> Quizás de todos los ***resultados*** de la revolución, ***el más importante*** fue la democracia del país. (Lorena 33–05)

In spoken language, in particular, logico-semantic relations such as cause and effect are more commonly realized by conjunctions. By contrast, this type of metaphor is called 'logical GM' because it involves what Martin (1993) calls 'buried reasoning', or the metaphorical realization of the logico-semantic relations (e.g., cause and effect) that, in a less metaphorical realization, would be expressed by conjunctions. This metaphorical realization of conjunctive relations by processes like 'resultar', 'causar', 'depender de' and nominal groups like 'los efectos', 'los resultados', 'las causas', 'las consecuencias', is particularly frequent in the written (academic) medium and in genres that explain and elaborate, such as the open-question essay and the research paper in the humanities. Explaining just such functionalities of GM, namely as a tool for organizing texts, is essential, so that students learn to handle the challenges presented by abstract text in the humanities and sciences.

In my analysis of the development of academic language in Latino students, the logical GM appear more frequently only after students start using other types of ideational GM. For example, Ana's introductory paragraph (presented above), which was written at the end of the academic year, presents several nominalizations as well as examples of logical GM. In a more congruent form these logico-semantic relations could have been expressed by means of a conjunction, as in

> El machismo aumentó **por** la guerra
> Los derechos de las mujeres se retrasaron **por** la guerra

Interpersonal grammatical metaphor

The last type of GM is the interpersonal GM. As stated, the main function of the ideational GM is to condense the information by way of packing more lexical items into one clause while deleting participants and the time of the processes; that is, the ideational GM is a more metaphorical way of expressing the meaning at the level of experience. The interpersonal GM, on the other hand, can be described as a metaphorical way to express interpersonal meanings that are congruently represented in mood and modality choices. The use of this kind of GM is especially important in academic language as it allows for a more explicit or implicit presence of the writer/speaker in the discourse. Expressions like 'creo, pienso, estoy convencida, estoy segura', known as 'explicitly subjective' (Martin 1997), express the modal assessment of ***probability*** in a clause that makes the speaker ***explicitly responsible for the assessment.***

> ***Yo pienso*** que toda la pelea fue inutil por que nada se mejoró con la guerra sino que con la comunicación que solo empezó mucho después que aya terminado la violencia. (Rosa 33–05)
>
> En resumen, ***yo estoy de la opinión*** de que Francisco Villa, Emiliano Zapata y otros no murieron en vano. (Lorena 33–05)
>
> ***En mi opinión, yo pienso*** que la revolución, aunque hubo muchas muertes de inocentes, sirvió mucho. (Lorena 33–05)

Modalization may also be made **explicitly objective**, through nominalizations of probability and usuality that construe the writer's presence and judgement either as a quality (adjective), 'es posible, probable, cierto, típico,' or as expressing a thing, like 'no hay posibilidad de . . .' The following are examples from the students' texts.

> Cuando se habla de la Revolución Mexicana, **es importante notar** cuáles fueron los beneficios de la guerra. (Ana 33–01)
>
> ***Es claro*** que el tema más importante del poema es la vida. (Ana 32–03)
>
> ***Además hay que tener en cuenta que*** el último verso de cada estrofa contiene la palabra 'nunca' . . . (Ana 32–03)
>
> ***Es evidente que*** desde el principio la autora se identifica con la gente de su país contándole de nuestra tierra. (Ana 32–03).

The use of SFL as a pedagogical framework will call for the explicit presentation of linguistic features that realize *objectiveness* or *subjectiveness* in the texts. Consequently, a clear understanding of the interpersonal GM is essential for students to become aware of the different lexicogrammatical resources they have to express meaning. Ana's examples of her final research paper (RP) and oral presention give us an idea of how students start working with these concepts. In the methodology section of the RP Ana, using an explicitly objective GM, writes:

> El proyecto actual: Sus intenciones y metodología
> ***La autora*** intentó llegar a una respuesta lacónica por medio de una investigación de 3 telenovelas de cada uno de los países con las más grandes sumas de telenovelas: México y Brasil. ***La investigación*** se enfocó en las décadas de los 1980s, 1990s y hasta lo más reciente de los años 2000. Casi 4.000 telenovelas fueron televisadas en México y Brasil, desde la década de los 80 hasta e año 2002 (Puga 1986, Cabrujas 2002, Fadul 1993). A principios de esta década, ocurrió una 'epifanía' mundial sobre la telenovela, con el incremento de su exportación como consecuencia de la necesidades de la globalización de los mercados latinoamericanos (Mazziotti 26b, Fonseca & Miranda-Ribiero 98, de Urbina & Lopez, 1999). Por esa razón, ***el enfoque de esta investigación*** se trató de las telenovelas transmitidas en los últimos 20 años en dos de los más grandes mercados de este género. . . . ***La autora*** reconoce que las telenovelas investigadas no fueron las más populares en los países que fueron transmitidas. Aún así ***la investigación*** logró acertar resultados concurrentes en las 6 telenovelas: La telenovela latinoamericana sirve para inculcar la idea patriarcal que la meta principal de la mujer es casarse, tener hijos y formar parte de una familia próspera. Además el matrimonio se describe como un ritualismo que no debería tomar lugar sin que las dos personas estén sumamente enamoradas de cada uno. (Ana 33–RP)

In other words, Ana is conveying objectivity by detaching herself from her work, for example, by using 'la autora' to refer to herself as the author of the work submitted. Even though the use of 'la autora' as authorial self-reference in academic register in Spanish in the humanities ultimately turns out not to be common or effective, it is clear that the student is experimenting with the interpersonal GM to present herself in a more objective way. On the other hand, in the oral presentation of her research paper she used subjective interpersonal GM in ways appropriate for the oral medium, showing that she is able to choose among different lexicogrammatical resources in academic Spanish in different environments of use:

> **Título**: La influencia de la telenovela en la cultura latinoamericana
> 1. ***Mi presentación*** es sobre las telenovelas
> 2. XXXar porque **tenemos** como diez minutos antes de que se termine la clase.
> 3. ahm . . . XXX
> 4. ***Lo que hice*** fue que . . . **recu** . . . ah . . . **hice** una . . .
> 5. una . . . investigación,

6. de todas las telenovelas {que salieron en México y Brasil desde 1980 hasta el año 2002}.
7. este: fueron como . . . ***voy a mencionar*** un total casi de cuatro mil . . . ah . . . telenovelas en total {que han salido desde . . . en estos veinte años}
8. ah . . . y: de esas ***escogí*** tres novelas de cada país . . . **son** XXX al resto de los resultados. (Ana 33–OP–03)

Conclusions and implications

From an SFL perspective, developing knowledge and understanding of the content area and developing control of the linguistic resources that construct and communicate that knowledge and understanding are essentially the same thing (Hasan 1996). The case of Spanish as language arts and a minority language in the context of education in the United States has its own characteristics and hence entails distinctive literate practices. In this paper I have focused on the use of GM in Spanish as a linguistic resource deployed by Spanish heritage speakers as a way of realizing academic language. Indeed, it is important to emphasize that GM in this context needs to be understood as a linguistic resource, i.e., a mechanism or process of the linguistic system (Derewianka: 2003) and not as a mere component of the language. GM is a way of meaning characteristic of the academy and professional contexts. For that reason students being apprenticed into the professions need to learn how to use language in the way their professional context finds meaningful, which presupposes that they know what ways of positioning are valued in what contexts in society.

There are numerous pedagogical implications of the use of GM in Spanish. Texts with a high degree of GM tend to be considered prestigious in Spanish-speaking cultures, as they are in English-speaking contexts. GM, particularly nominalization, is a typical feature of many types of written (academic) texts and is usually associated with the notions of 'abstraction' and 'distance' in the humanities and social sciences and with technicality in the sciences (Halliday 1993). Knowing how to use GM in academic registers is an essential part of developing academic language. Realizing the similarities and differences between Spanish and English can help bilingual students to transfer these features from one language into the other.

Much more research is needed to identify academic language development, but SFL provides a framework for explaining how lexicogrammatical features mean in the academic context. The analysis discussed here indicates that communicatively effective features of Spanish academic texts are indeed identifiable and amenable to specification. Further research into the nature of the pedagogical effectiveness of a GM-informed approach to texts with different student groups for a variety of purposes could then lead to an informed agenda for the curriculum in Spanish as a heritage language in the United States.

Notes

1 According to the U.S. Census Bureau, Hispanics (the term assigned by the U.S. Census and referring to people of all ethnic backgrounds but whose origin is a Spanish-speaking country) are the fastest growing segment of the population, totalling 37.4 million in March 2002 and the largest minority in the United States. Half of all Latinos live in just two states: California and Texas. Latinos in California accounted for 11.0 million persons and 31 per cent of the Hispanic population in the United States, while Texas has 6.7 million persons, that is, 19 per cent. The number of Latino-owned firms has grown immensely in the last ten years, with a figure of 1,574,159 being reported in the last census.

2 With regard to their sociocultural background, most of the Latino students at the University of California, Davis, are second- or third-generation Spanish speakers who are the first in their families to access higher education. This program aims at developing their academic proficiency in oral and written modes. When entering the program, students bring with them the oral features of Spanish of interpersonal communication and informal conversational registers; over the course of the year of instruction they move along the continuum of language, developing some features of academic language.

3 The corpus of written and oral texts was studied following a genre/register analysis of genre (text type) and its functional components to identify the appropriateness and effectiveness of the students' texts according to the purpose and context of the situation. Then an SFL clause combining analysis, in combination with lexical density and nominal density, was applied to the corpus to determine the grammatical intricacy and lexical density of the texts. The findings of this analysis help explain students' movement along the continuum of language development in Spanish. For further information on the analysis of the corpus, see Colombi (2002).

4 All names are pseudonyms to protect students' identity.

5 All examples come from the first version of their multiple version assignments and have been copied literally without editing or correction.

6 The oral presentation is a genre that falls within the category of public speech and forms part of a continuum of genres of academic language. It is spoken language, inasmuch as the interlocutors are co-present in the realization of the text; however, it is not spontaneous because students have researched and composed it in writing ahead of the presentation. In the cases analysed, the students presented a written outline on the day of their presentation, followed by the research paper with a total of three versions.

7 This segment belongs to the conclusions section of the oral presentation.

8 This excerpt comes from the development of Ana's essay, 'Las reformas dentro de la Revolución Mexicana', quoted above.

References

Acevedo, R. (2003) 'Navegando a través del registro formal', in A. Roca and M. C. Colombi (eds), pp. 257–68.

Achugar, M. (2003) 'Academic registers in Spanish in the U.S.: a study of oral texts produced by bilingual speakers in a university graduate program', in A. Roca and M. C. Colombi (eds), pp. 213–34.

August, D. and Hakuta, K. (eds) (1997) *Improving Schooling for Language-Minority Children.* Washington, DC: National Academy Press.

August, D. and Hakuta, K. (eds) (1998) *Educating Language-Minority Children.* Washington, DC: National Academy Press.

Belcher, D. and Connor, U. (eds) (2001) *Reflections on Multiliterate Lives.* Clevedon, England: Multilingual Matters.

Beykont, Z. (ed.) (2002) *The Power of Culture: Teaching across Language Differences.* Cambridge, MA: Harvard Education Publishing Group.

Carreira, M. (2003) 'Profiles of SNS students in the twenty-first century', in A. Roca and M. C. Colombi (eds), pp. 51–77.

Cenoz, J. and Genesee, F. (eds) (1998) *Beyond Bilingualism: Multilingualism and Multilingual Education.* Clevedon, England: Cromwell Press.

Chafe, W. and Danielewics, D. (1987) *Properties of Spoken and Written Language.* Berkeley, CA: Center for the Study of Writing.

Christie, F. (2002a) *Classroom Discourse Analysis.* London: Continuum.

Christie, F. (2002b) 'The development of abstraction in adolescence in subject English', in M. J. Schleppegrell and M. C. Colombi (eds), pp. 45–66.

Colombi, M. C. (1997) 'Perfil del discurso escrito: teoría y práctica', in M. C. Colombi and F. J. Alarcón (eds), *La enseñanza del español a hispanoblantes: praxis y teoría.* Lexington, MA: D. C. Heath, pp. 175–89.

Colombi, M. C. (2000) 'En vías del desarrollo del lenguaje académico en español en hablantes nativos de español en los Estados Unidos', in A. Roca (ed.), *Research on Spanish in the United States.* Somerville, MA: Cascadilla Press, pp. 296–309.

Colombi, M. C. (2002) 'Academic language development in Latino Students' writing in Spanish', in M. J. Schleppegrell and M. C. Colombi (eds), pp. 67–86.

Colombi, M. C. (2003) 'Un enfoque funcional para la enseñanza del lenguaje expositivo', in A. Roca and M. C. Colombi (eds), pp. 78–95.

Colombi, M. C., Pellettieri, J. L. and Rodríguez, M. I. (2001) *Palabra abierta.* Boston: Houghton Mifflin.

Cummins, J. (2000) *Language, Power and Pedagogy. Bilingual Children in the Crossfire.* Clevedon, England: Multilingual Matters.

Derewianka, B. (2003) 'Grammatical metaphor in the transition to adolescence', in A. M. Simon-Vandenbergen *et al.*, pp. 185–219.

Eggins, S., Wignell, P. and Martin, J. R. (1993) 'The discourse of history: distancing the recoverable past', in M. Ghadessy (ed.), *Register Analysis: Theory and Practice.* London: Pinter, pp. 75–109.

Fuentes, C., quoted in M. C. Colombi *et al.*, 2001.

Gibbons, J. (1999) 'Register aspects of literacy in Spanish'. *Written Language and Literacy*, 2, 63–88.

Halliday, M. A. K. (1985) *Spoken and Written Language.* Geelong, Australia: Deakin University Press; (re-edited in 1989) Oxford: Oxford University Press.

Halliday, M. A. K. (1993) 'Towards a language-based theory of learning', *Linguistics and Education*, 5, 93–116.

Halliday, M. A. K. (1994) *An Introduction to Functional Grammar (2nd edn).* London: Edward Arnold.

Halliday, M. A. K. (1998) 'Things and relations: regrammaticising experience as technical knowledge', in J. R. Martin and R. Veel (eds), *Reading Science: Critical and Functional Perspectives on Discourses of Science.* London: Routledge, pp. 185–235.

Harley, B., Patrick, A., Cummins, J. and Swain, M. (eds) (1990) *The Development of Second Language Proficiency.* Cambridge: Cambridge University Press.

Hasan, R. (1996) 'Literacy, everyday talk and society', in R. Hasan and G. Williams (eds), *Literacy and Society*. London: Longman, pp. 377–424

Jones, J. (1990) 'Grammatical metaphor and technicality in academic writing: an exploration of ESL (English as a second language) and NS (native speaker) student texts', in F. Christie (ed.), *Literacy in Social Processes*. Deakin University, Australia: Centre for Studies of Language in Education, pp. 178–98.

Martin, J. R. (1993) 'Life as a noun: arresting the universe in science and humanities', in M. A. K. Halliday and J. R. Martin (eds), *Writing Science: Literacy and Discursive Power*. Pittsburgh, PA: University of Pittsburgh Press, pp. 221–67.

Martin, J. R. (1996) 'Waves of abstraction: organizing exposition'. *The Journal of TESOL France*, 3, 87–104.

Martin, J. R. (1997) 'Analysing genre: functional parameters', in F. Christie and J. R. Martin (eds), *Genre and Institutions: Social Processes in the Workplace and School*. London: Cassell, pp. 3–39.

Martínez, G. (2003) 'Negative Transfer of Literacy Skills: pedagogical considerations for writing instruction in the SHL classroom'. Presentation given at LASSO, 17–19 October 2003. The University of Texas-Pan American, Endinburg.

Martín-Rodríguez, M., quoted in M. Cecilia Colombi *et al.*, 2001.

Ravelli, L. (1988) 'Grammatical metaphor: an initial analysis', in E. H. Steiner and R. Veltman (eds), *Pragmatics, Discourse and Text: Some Systemically-inspired Approaches*. London: Pinter, pp. 133–47.

Ravelli, L. and Ellis, R. (eds) (2004) *Analysing Academic Writing: Contextualized Frameworks*. London: Continuum.

Roca, A. and Colombi, M. C. (eds) (2003) *Mi lengua: Spanish as a Heritage Language in the United States*. Washington, DC: Georgetown University Press.

Schleppegrell, M. J. (2004) *The Language of Schooling*. Mahwah, NJ: Lawrence Erlbaum.

Schleppegrell, M. J. and Colombi, M. C. (1997) 'Text organization by bilingual writers'. *Written Communication*, 14, 481–503.

Schleppegrell, M. J. and Colombi, M. C. (eds) (2002) *Developing Advanced Literacy in First and Second Languages*. Mahwah, NJ: Lawrence Erlbaum.

Simon-Vandenbergen, A., Taverniers, M. and Ravelli, L. (eds) (2003) *Grammatical Metaphor*. Amsterdam/Philadelphia: John Benjamins.

Unsworth, L. (ed.) (2000) *Researching Language in Schools and Communities*. London: Cassell.

Valdés, G. (2001) *Learning and Not Learning English*. New York: Teachers College Press.

Valdés, G. (2003) *Expanding Definitions of Giftedness: Young Interpreters of Immigrant Background*. Mahwah, NJ: Lawrence Erlbaum.

Valdés, G. and Geoffrion-Vinci, M. (1998) 'Chicano Spanish: the problem of the "underdeveloped" code in bilingual repertoires'. *Modern Language Journal*, 82, 473–501.

8 Creating textual worlds in advanced learner writing: the role of complex theme

Marianna V. Ryshina-Pankova

Advanced language ability in the native and a foreign language has been associated with expansion of registers, which includes the acquisition of genres representing various institutional, educational and professional settings and comprising secondary discourses of public life (Byrnes and Sprang 2004; Gee 1998; Matthiessen, this volume). Such contexts are often dominated by written communication that itself is characterized by a distance between the writer and the social process and between the writer and the audience. This detachment places special demands on the language used in written genres. Unlike in many oral varieties, where language enacts or accompanies a social process, language in the written mode construes social reality and reflects on it. Furthermore, the distance between writer and addressee in written communication does not allow the writer to receive or react to the reader's immediate feedback as is possible, for example, in a dialogue, where turn-taking enables the addressee to participate in the communicative event. Among other things, that distance requires writers to provide the right amount of background information to their readers, whom they generally do not know personally, in order to anticipate their questions or concerns, and to express their own positions and attitudes with regard to the issue being discussed.

The role that language plays in such institutional contexts of schooling or the professions has a direct impact on patterns of language use. Language used as reflection and as constituent of social processes differs from language as action in terms of its lexical density, grammatical complexity and discourse organization. In particular, encoding reflection on and evaluation of reality necessitates objectification of the dynamic nature of reality, a process that has evolved especially under the demands of science to 'hold the world still – to stop it wriggling, so to speak – in order to observe and study it' (Halliday 1991: 10).

Regarding the lexicogrammatical aspects, this occurs through nominalization, whereby verbs as processes, adjectives as descriptions or adverbs as circumstances are turned into nouns or things that can be further described, classified and organized in terms of various logical relations (Halliday and

Martin 1993; Colombi and also Schleppegrell, this volume). On the discourse level, academic or professional written texts require careful planning so as to be able to render information successfully or, more generally, to achieve the writer's communicative goals by providing readers with sufficient contextualization of the matter in question and guiding them through the stages of the text.

The extent to which that ability to recreate reality textually manifests itself in written secondary discourses produced by foreign language writers is the focus of this paper. It examines advanced foreign language writing by utilizing the understandings within systemic-functional linguistics (SFL) for *coherence* as appropriateness to the communicative purposes of a specific situation and to a larger cultural context in which a piece of writing is produced and for *cohesion* as internal unity of texts. The study it reports on examines coherence with the social context and cohesion within text in terms of information-structuring patterns. Specifically, organization of meanings by means of textual resources in learner essays is investigated through the constructs of *textual stages* or *moves* and *theme selection* within those stages.

Coherence and cohesion in text linguistics, SFL and genre theory

Research in text linguistics and discourse analysis has long differentiated between the constructs of coherence and cohesion. In general, coherence was understood as a non-surface realization of connectedness that resides in implicit semantic relations between propositions, in information structure and readers' background knowledge frames. As an *implicit* phenomenon it was investigated in terms of (1) relations between cohesive ties as in Hasan's cohesive harmony theory (Halliday and Hasan 1989), (2) thematic progressions as in Daneš' (1974) framework, (3) topical development as in Lautamatti's approach (1987, 1990), or (4) semantic relations between propositions as elaborated by Winterowd (1970), Fahnestock (1983) and also Halliday and Hasan (1976). By comparison cohesion was defined by means of the surface-level resources that realize coherence, such as cohesive ties. An influential taxonomy of such explicit cohesion features, which include reference, substitution, ellipsis, lexical cohesion and conjunctions, was developed by Halliday and Hasan (1976) and further applied in various studies that attempted to relate the number or type of cohesive ties to the overall well-formedness of texts. However, the majority of such studies demonstrated only low correlation between the quantity of cohesive ties and overall coherence and success of texts (Carrell 1982; Connor 1984; Khalil 1989; Tierney and Mosenthal 1981; Witte and Faigley 1981).

In all these cases, research into the analysis of textual unity was based on purely textual features, implicit or explicit. However, in order to account for well-formedness of texts as actions in a communication process that aims at realizing specific communicative purposes typical of specific social contexts, a contextually-based approach to coherence is needed. SFL and genre theory, as developed by the Australian linguist Halliday (1989, 1994) and his followers, particularly Christie (1985, 1986), Hasan (1995, 1996), Martin (1985, 1997, 1998, 1999), Martin and Rose (2003), Ventola (1991) and Ventola and Mauranen (1996), provides one model for analysing how a particular instance of language use is coherent with the social context in which it occurs, as well as cohesive within itself.

To accomplish that goal, SFL and genre theory define contextual coherence as coherence of genre and coherence of register. In line with Martin (1984), genres are 'staged goal-oriented social processes' that represent more or less stable or typical ways of achieving communicative goals in a particular social situation. Generic coherence refers to a predictable culturally typical sequence of obligatory, optional and recursive stages through which a communicative purpose specific for a particular verbal action (genre) is gradually realized. In genre theory, then, a text is considered coherent and complete when, in order to achieve its communicative goal, it moves through all the obligatory stages of the structure potential of the genre (Hasan 1996). For example, the following genre stages make the genre of recipe generically coherent and successful (Eggins 1994: 44): title, enticement, ingredients, method and serving quantity.

Register coherence refers to the possible combinations and their realization in text of three contextual variables singled out as most important by SFL for understanding language in context. These variables are *field* – the nature of social processes and subject matter, *tenor* – the relationship between the participants and *mode* – the role of language in the instance of communication. According to Halliday, each of the contextual variables corresponds to a particular type of meanings human beings make in response to social functions and needs that arise in certain speech situations and that language users wish to realize. *Field* corresponds to experiential meanings concerned with representation of experience, *tenor* is related to interpersonal meanings concerned with the relationship between participants in interaction and *mode* corresponds to textual meanings concerned with organization of the experiential and interpersonal meanings.

Within mode, one of the most powerful language systems responsible for organizing interpersonal and ideational meanings into genre-specific textual stages is the system of theme. As the first element in the clause (Halliday 1994) and a point of departure for the text, theme foregrounds linguistic realizations of field or tenor by means of particular patterns of theme selection and progression. By prioritizing certain interpersonal and experiential meanings, theme helps organize text so as to achieve the communicative goals of particular contexts, thus acting as an instrument for contextual generic coherence (Drury 1991; Francis 1989; Fries 1994; Ghadessy 1995; Kuo 1995; Lotfipour-Saedi and Rezai-Tajani 1996; Nwogu and Bloor 1991). At the same time, theme operates on the local level as a textual connecter between the previous and following discourse, thereby enabling internal cohesion of texts (Mauranen 1996; Schneider and Connor 1990; Witte 1983). The current study builds on this research and investigates how theme as a resource for contextually coherent and textually cohesive organization of ideational meanings is used by advanced learners of German.

Data base and curricular framework of the study

Data for the study were collected from students representing Levels 3 (23 essays), 4 (14 essays) and 5 (18 essays) of the curriculum in the German Department at Georgetown University. These levels constitute the upper or advanced levels of this content-oriented, task-based and genre-informed curriculum. Together they aim at a steady progression from personal narrative writing toward increasingly more elaborated public and genre-specific forms of writing. Instruction at Level 3 is geared towards developing student abilities for constructing discourse that expands personal stance into the public sphere mainly through comparison and contrast, cause and effect and presentation of alternative perspectives. Major discourse patterns beyond the chronological narrative are taught at this level. Level 4 further extends personal discourses into the realm of the abstract by focusing on the secondary discourses of public life. At Level 5, students continue to refine their discursive abilities in academic writing by concentrating on specific genres, like literary criticism or the research paper.

For the writing task that provided the data students were prompted to imagine themselves as international exchange students at a German-language university who had been asked to submit a contribution to an 'International Readers' Corner' (*Internationale Leseecke*) column in the student newspaper that features reviews of books international students would recommend to their German peers. Either a fictional or a non-fictional book could be used. At minimum, the review was to include a summary of the storyline or major thematic points of the book, reasons why the book was compelling and insightful, an argument for its suitability for the German student audience and appropriate information about the author and his/her significance.

The genre of book review was selected for three reasons. First, book reviews exemplify academic writing that is highly valued in the educational/scientific environment (Hyland 2000). Second, it was hypothesized that book reviews would allow students to show their writing abilities in the construction of various functional environments, such as description, narration, evaluation and argumentation, all present within this genre. A third rationale for choosing the book review was its authenticity as a genre that students do use throughout their academic career.

The data were later complemented by book reviews written by native speakers of German who were students at the Freie Universität in Berlin and at the University of Trier (ten essays). In addition, native speaker data from a German Internet literature magazine *Literaturzirkel, Belletristik, Science Fiction, und Fantasy*, http://www.ma1stnews.de/literaturzirkel/ were included (20 texts). These texts provided information about native-speaker production of the genre and the opportunity to compare and contrast it with L2 learner writing.

Determining the moves structure of book reviews

To determine how theme organizes student book reviews, a moves structure as a staged realization of their contextual communicative purposes was first established, based on the analysis of all texts. The following five moves and their roles were identified:

I Motivation – optional, ordered (always first)
II Content – optional or obligatory when comment is not present, unordered, repeated
III Comment – optional or obligatory when content is not present, unordered, repeated
IV Author – optional, unordered
V Evaluation – obligatory, unordered, repeated

Table 8.1 more closely describes the meaning and purposes of these moves.

According to this analysis, the genre is characterized by two obligatory moves combinations, either Content and Evaluation or Comment and Evaluation. Only Content, Evaluation and Motivation are important for the

Table 8.1 Communicative purposes of the moves of the genre '*Buchbesprechung/Buchempfehlung*' (book review, book recommendation)

Moves and their functions	Lead questions to identify the moves
I. **Motivation** – to motivate the choice of a particular book	*Why choose this particular book? What makes it special?*
II. **Content** – to describe major themes of the book, to narrate some aspects of the content	*What is the book about? What happens?*
III. **Comment** – to interpret the content of the book, to show how the author presents the content	*What does the content mean? How does the author present the content? What are the author's preferred topics/genre/style?*
IV. **Author** – to present autobiographical information on the author	*Where and when was the author born? Where did the author live? What makes her/him famous/special?*
V. **Evaluation** – to justify the choice of the book by describing the book's aesthetic qualities with regard to its language, plot, characters, by stating its emotional, aesthetic and intellectual effect on the reader	*What kind of book is it, e.g., fun, well/clearly written? What effect does it have on the reader, e.g., pleasant/interesting to read, full of suspense? What can one learn from it? Why is it suitable for a particular audience? Is it worth reading? How does it compare to similar books? Why would one recommend this book?*

discussion in this paper and are described below. The Content move lists major themes of the book and narrates aspects of its content in order to inform the reader about the happenings in the book.[1] The listing of themes is not very elaborate and renders the subject matter of the book without providing much interpretation. The following example demonstrates how the Content move often includes one or two sentences that summarize the plot in general terms, e.g., *es handelt vom, es geht um* . . . (it deals with, it pertains to . . .) before narrating about the actual events in the story:

3.h.S3.2082.b.cha[2]
Content
*T: es geht direkt um das leben von englischen kolonisten in burma und ihre erfahrungen, während in parallel es ein kritik von imperialismus und kolonisation ist.

The Content move most often follows the typical narrative structure detailed by Labov (1972): listing of general themes as an abstract of the story, complicating action and resolution.

The Evaluation move is necessary to persuade the reader to select the book by describing its aesthetic qualities with regard to language, plot and characters, and by stating its emotional, aesthetic and intellectual effect on the reader. It also directly appeals to the reader to read the book. In the example below, the writer of the book review evaluates the book by identifying its effect on the reader, expressing satisfaction with the book, and recommending it as a pleasant read.

ns6.b.cha
Evaluation
impact
*T: die zunehmende rührung beim lesen, schafft eine tiefe verbundenheit zu diesem jungen mädchen.

satisfaction
*T: das lesen dieser briefe ist das reinste vergnügen.
*T: denn die junge autorin versprüht nicht nur treffenden witz, beschreibt in ihrer gewissen naivität gefühle, die jedem leser nur allzu bekannt sind, so daß wir uns intensiv mit der jungen protagonistin identifizieren können.

recommendation
*T: um das lesevergnügen zum äußersten zu steigern empfehle ich, es an einem besonders schönen sommertag während eines picknickes in einem park oder an einem see zu lesen, am besten zusammen mit der musikalischen untermalung von johann strauß' walzern.

Finally, the Motivation move, when present, always starts the book review. It enables the writer to motivate the reader to choose a particular book by staging an argument that establishes a cause for reading the book, explaining the circumstances that led to its being chosen, positively evaluating it by pointing out its unique qualities, or even voicing an explicit appeal to read the book. The Motivation stage also functions as an eye catcher, that is, as a text that aims to attract the reader to continue reading the book review. The following example demonstrates how, through a problem-solution structure, (problem: *Sie haben Angst vor dicken Romanen, keine Ahnung, welches Buch weder nicht zu schwer noch lesewert wäre – you are afraid of fat novels, have no idea which book would not be too difficult and yet worth reading;* solution: *wäre sein Buch die beste Wahl – this book would be the best solution*)[3] the writer makes the case for his book. At the same time, the reader is attracted to the text by means of the direct dialogue structure employed by the writer in the review text itself.

h.5.S3.2020.cha
Motivation
*T: wie oft haben sie gedacht, dass sie etwas mehr ueber die klassische literatur des vergangenen jahrhunderts wissen sollen?
*T: aber haben angst vor dicken romanen, die in einem fuer sie wirklich unverstehbaren stil geschreiben werden?
*T: wie oft, dann, haben sie keine ahnung, welches buch weder nicht zu schwer noch lesewert waere?
*T: wenn sie wie ich solche schwierigkeiten begegnen haben, vielleicht koennte ich mit ihnen meine erfahrung mit so einem ausgezeichneten buch teilen, die sie hoffentlich zum lesen dieser position einladen wuerde.
*T: ueber den autor haben zweifellos alle gehoert: franz kafka.
*T: na ja, seine geschichte gar keine lustigen erzaehlungen sind.
*T: aber wenn sie manchmal serioes fuehlen, waere sein buch 'die verwandlung' die beste wahl.

Realization of ideational themes

The second step in the analysis was identification of ideational theme. In line with Halliday (1994), it was operationalized as the first element in a clause with a transitivity function, such as various types of participants (e.g., *actor, senser*) or processes. Ideational theme was only marked for main clauses or for dependent clauses if they took the first position in a sentence. I hypothesized that an ideational theme that contains several elements plays a special role in bringing about coherence and cohesion of moves by packing information relevant to a specific move and to the movement of information inside a move. In line with this hypothesis, I expected essays written by learners at more advanced curricular level or essays written by native speakers to contain more themes that are complex. To examine the hypothesis, ideational themes that consisted of a grammatical clause, more than three lexical elements, or a grammatical metaphor were identified and marked respectively as *grammatically intricate, lexically complex* and *lexically dense*. The remainder of the paper demonstrates how the first two types of these complex themes were used by advanced language learners as coherence- and cohesion-building devices in selected moves of their book reviews.

Grammatically intricate themes

Grammatical intricacy in themes was identified as thematization of subordinate clauses that include (1) various circumstantial clauses of time, condition, concession and reason, (2) modifying clauses, (3) thematic equatives as distinguished by Halliday and Matthiessen (2004) and (4) other types of nominalized clauses.

1) *T:**als der junge in der grundschule war $COMPclause**, ist er sehr krank geworden und durfte nicht in der schule gehen (*when the boy was in grade school. . .*).
2) *T:**ohne seine humanistischen wurzeln zu verleugnen $COMPclause**, entwickelt sich der text aus der zeittypischen tendenz zur satire über stände, charaktere und menschliche schwächen (*without denying his humanistic roots . . .*).
3) *T:**was sie letztendlich zur tat treibt $COMPclause** sind die gleichtönigkeit und leere dieses daseins (*what finally drives them into action . . .*).
4) *T:**dass die schwestern sich nicht innerhalb zwei wochen im jeden bezug verstehen $COMPclause** beweisst wie ehrlich sie zu einander sind (*the fact that the sisters do not, within the two week time period, get along with each other in every respect . . .*).

As Table 8.2 indicates, a significant increase in the use of thematized subordinate clauses can be observed from Levels 3 and 4 to Level 5, suggesting a development in text-constructing strategies.

As a cohesion instrument, subordinate clauses are an excellent resource for establishing connections within moves. The information they contain not only restates the previous text but, being placed in thematic position, it also serves as a framework for the following discourse. As a coherence strategy, thematization of subordinate clauses of various types enables writers to render the content of the book in a compact way by establishing logical relations, particularly through circumstantial clauses, between events and protagonists' actions, and to introduce argumentative structures into the book review in order to persuade the audience to read the book.

In that context it is curious to note the decrease in clausal themes in NS book reviews. It can be explained in terms of a preference for another organizational strategy, namely the use of lexically complex and lexically dense themes (grammatical metaphors), which can play a similar role in structuring texts as do clausal themes.

Table 8.2 Clausal themes across levels

	NS	**Level 5**	**Level 4**	**Level 3**
Clausal themes	6.4%	9.6%	6%	6.5%

Clausal themes in Evaluation in NNS texts

The role of clausal themes in creating coherence while also serving a cohesion-building function can be illustrated particularly well within the Motivation and Evaluation moves. The most frequent type of clausal themes for organizing these moves are conditionals (61.5% of all clausal themes at Level 5, 33.3% at Level 4 and 50% at Level 3). Thematization of conditional clauses in moves whose communicative goal is to convincingly argue for the choice of the book is not accidental. Conditional clauses have been shown to play a prominent role in structuring argumentative discourse. Schiffrin (1992), for example, describes how conditional sentences, through the *protasis* (the 'if' part of the clause), contribute to construction of an argument by creating a relation with what she calls 'message-level topics'. This relation is manifested in the form of a summary of the previous points that is brought about by means of repetition, reformulation or use of inferrables that all refer back to the previous text. In other words, by relating to message-level topics conditionals can be understood as a *cohesive* function. At the same time, as Schiffrin demonstrates, the second part of the conditional sentence, the *apodosis*, draws a consequence from the information provided in the protasis, thus supplying evidence for 'the speaker topic' or the overall goal of the speaker/writer discourse that ultimately defines its global *coherence* structure.

The Evaluation and Motivation moves in NNS data show how conditional clauses enable writers to stage their argument in such a way as to provide a better motivation to read the book and, at the same time, to act as local textual connectors. In the following Level 5 example, one can see how, in line with Schiffrin (1992), the protasis restates all the previous arguments for the choice of the book, while the apodosis presents a conclusion based on these arguments, which supports the global point of the text.

5.h.S3.2029.b.cha; line 48.
Evaluation

*T: **aus der perspektive der sprache ist dies buch ein gutes beispiel von englisch in seiner schoensten form (1).**

*T: fitzgerald hat das talent ereignisse accurate und zur gleichen zeit poetisch zu beschreiben.

*T: am eindruckfollsten ist eine szene in der gatsby eine party hat.

*T: **die beschreibungen von der atmosphaere, das essen, die kleider usw gibt eine spezifischen eindruck raus (2).**

*T: in seiner beschreibung zeigt fitzgerald nicht nur wie alles aussieht, aber gibt den leser auch indirekte hinweise auf was wirklich sich hier vorspielt.

*T: wenn sie sich entscheiden dieses buch zu lesen nemmen sie acht auf die benutzung der farbe.

*T: es gibt so viele tiefere andeutungen in diesem buch, die machen **es viel spass zu lesen (3).**

*T: **wenn sie ein buch suchen das gut geschrieben ist (1), interessant ist (3), und einen blick auf amerikanische kultur bietet (2) $CONDITION**, lesen sie 'the great gatsby' von f scott fitzgerald.

Summarizing the positive characteristics of the book in the protasis that had been explicated in the previous discourse – (1) it is an example of written English in its most beautiful form, (2) familiarizes the reader with a certain cultural atmosphere through descriptions of food and clothing, and (3) is an enjoyable read – creates solid evidence for justifying the value of the book in the apodosis and enables the author to formulate a strong conclusive recommendation (in the form of an imperative) to read it.

The same cohesion and coherence creating function of conditionals is manifested in the construction of the following Motivation move of another Level 5 learner:

5.h.S3.2020.b.cha
Motivation
*T: wie oft haben sie gedacht, dass sie etwas mehr ueber die klassische literatur des vergangenen jahrhunderts wissen sollen.
*T: aber sie haben angst vor dicken romanen, die in einem fuer sie wirklich unverstehbaren stil geschriben werden?
*T: wie oft, dann, haben sie keine ahnung, welches buch weder nicht zu schwer noch lesewert waere?
*T: **wenn sie wie ich solche schwierigkeiten begegnen haben $CONDITION**, vielleicht koennte ich mit ihnen meine erfahrung mit so einem ausgezeichneten buch teilen, die sie hoffentlich zum lesen dieser position einladen wuerde.
*T: ueber den autor haben zweifellos alle gehoert: franz kafka.
*T: na ja, seine geschichte gar keine lustigen erzaehlungen sind.
*T: **aber wenn sie manchmal serioes fuehlen $CONDITION**, waere sein buch 'die verwandlung' die beste wahl.

Here, the protasis in the first conditional summarizes the diverse points of the argument (problems with the choice of the book), while the apodosis draws a justified conclusion from the argument: it constitutes an outstanding book that provides a solution to the problems. The second conditional clause in this Motivation functions in the same way for the second previously identified problem. It connects to the idea that stories are serious by expanding it causatively: serious stories require a serious reader. Thematization of the serious reader in the protasis allows the author of this book review to create a presupposition that such a reader actually exists and base her conclusion about the value of the book on this presupposition. This feature of thematized conditionals to present information as given, presupposed or inferrable (Schiffrin 1992) allows writers to lead or manipulate the reader to accept their position:

> If something is presented to us as a topic, we are inclined to accept it as 'shared' information even though it may be quite new to us. We may then take the further step of accepting it not only as 'shared' but also as 'true' (Schmid 1999: 79, cited after Brömser 1984: 343).

Similarly, Haiman argues that the protases of conditional clauses are presupposed to be true and are thus 'immune to challenge or denial' (1986:

218). In the case of book reviews, activated presuppositions of existence in the protases are used to convince the reader that the book is worth reading – even without much preparatory argumentative grounding. The following Evaluation move from yet another Level 5 book review illustrates the point.

5.h.S3.1035.b.cha; line 57.
Evaluation
*T: **wenn man niemals vorher etwas von douglas adams gelesen hat $CONDITION**, soll man mit 'per anhalter durch die galaxis' anfangen.
*T: das ist ein kurzes buch, in dem adams eine komische geschichte erzählt.
*T: **wenn man adams seltsamer sinn für humor überlegen kann $CONDITION**, soll man die folgende bucher auch ausprobieren.
*T: und wer mehr von douglas adams noch will, soll 'dirk gentlys holistische detektivagentur' auch annhemen.
*T: es macht immer spass, ein douglas adams buch zu lesen.
und wenn man noch nicht angefangen ist, diese werke zu genießen $CONDITION, soll man sofort 'per anhalter durch die galaxis' aufheben.
*T: und erinnern sie sich daran, 'keine panik'!

The second and third conditional protases in this example presuppose that there are people who will understand the author's peculiar humour and readers who will enjoy the author's works. On the bases of these protases, pleas for reading the book are made.

Clausal themes in Evaluation and Motivation in NS texts

While conditional clause themes are quite prominent in the Evaluation and Motivation move in NNS book reviews, they never occur in the Evaluation move and only once in the Motivation move in NS texts. Here, the most frequent type of clausal themes in the Evaluation move are nominalized clauses that also include thematic equatives. They constitute 75% of all clausal themes used in the NS Evaluation move. Particularly interesting is that these clauses display similar semantics of condition-consequence that are also important for staging the argument in support of the book in NNS texts.

7.ChasmCity.cha: line 64. Evaluation
*T: **wer hervorragende science fiction sucht**, sollte den roman jedoch im original lesen! (*anyone who is looking for superb science fiction . . .*)

7.DieSpurdesSelcti.cha: line 54. Evaluation
*T: **wer dieser fährte folgt**, wird mit bester unterhaltung belohnt! (*anyone who follows that trail . . .*)

ns3.b.cha: line 97. Evaluation
*T: **wer also an kognitiv anspruchsvoller, aber trotzdem sprachlich hochwertiger literatur interessiert ist**, sollte das buch in jedem fall lesen. (*anyone who is interested in intellectually demanding, yet highly valued literature from the standpoint of language use . . .*)

Structurally, however, they can be considered as more complex and compact due to their status as noun substitutes. Interpersonally, they display a stronger persuasive force: due to the exclusive thematization of the reader (*wer*) they appeal to the audience in a more direct way than do clauses that thematize both condition (*wenn*) and a reader.

The resource of nominalized clausal themes available to NS is exploited in their book reviews in yet other ways to structure the Evaluation move. They contribute to coherence of the Evaluation move by constituting relational clauses whose theme objectifies an aspect of reality and necessitates or enables an evaluation or description of this aspect in the rheme:

7.OliviaJoules.cha: line 66. Evaluation
*T: **ob dieser sprachduktus der charakterisierung dienen soll**, ist leider **nicht erkennbar**. (*whether this form of language use is intended to serve character depiction . . .*)

ns2.b.cha: line 79. Evaluation
*T: **ob sieben tage immer dafür ausreichen, wie coelho im letzten satz seines nachwortes anmerkt**, ist sicherlich **fraglich**. (*whether seven days always suffice for that purpose, as coelho suggests in the final sentence of his epilogue . . .*)

ns7.b.cha: line 50. Evaluation
*T: **in diese kürze mehr hineinzulesen** ist meiner meinung nach **kontraproduktiv**. (*to read more into this brevity . . .*)

In NNS texts, nominalized clausal themes appear as well, but constitute only a very small number of occurrences. Nevertheless, one can observe an increase in use from Level 3 to Level 5, as is evident from Table 8.3.

An increase at Level 5 suggests a gradual appropriation by more advanced NNS writers of specifically German discourse structures for this genre. This is noteworthy, inasmuch as fronting of nominalized clauses is not a common discourse strategy in English because of the restrictions on elements that occupy the subject position in the English clause (Steiner and Ramm 1995), as the above translation of sample ns.2.b indicates. In fact, parallel constructions in English can be considered clumsy and non-native. In German, on the other hand, they present a powerful resource that enables writers both to structure the move on the global level and to connect locally.

Lexical complexity in theme

Lexical complexity in theme was identified with regard to the number of

Table 8.3 Nominalized clausal themes in NNS texts

Level 5	Level 4	Level 3
14.3%	4.5%	4.3%

ideational lexical elements in a theme (excluding the clausal themes that were considered grammatically intricate). Only themes that contained more than three ideational lexical items were coded as lexically complex. Articles, possessive and demonstrative pronouns, as well as coordinated constituents, were not counted as separate lexical items.

Lexical complexity comes about as a result of modifications of the noun phrase. In structural terms, these modifications can be quite varied. Observed pre-nominal modification occurred by means of (1) adjectives, (2) extended attributes, (3) prepositional phrases and (4) pre-nominal participial constructions. Post-nominal modification was identified when (5) appositions, (6) participials and (7) embeddings through relative clauses were used.

(1)32.m.S3.2067.b.cha: line 23.

*T: **und das *wichtigste* buch von dieser genre** ist 'burning chrome', eine gruppe der kurzen geschichten von william gibson. (*and the most important book of this genre . . .*)

(2)5.h.s3.2020.b.cha: line 35.

*T: ***gegen den surrealistichen hintergrund gesetzte* geschichte** faengt irgendwo, als eines tages gregor samsa, der durchschnittliche buero-arbeiter, wachst sich in seinem bett als ein riesiges ungeziefer! (*the story set against this surreal background . . .*)

(3)5.m.S3.1027.b.cha: line 48.

*T: **das verstehen *von globalisation und ihre auswirkungen*** ist notwendig fuer alle einwohner der welt. (*an understanding of globalization and its consequences . . .*)

(4)4.h.S3.2008.b.cha: line 11.

*T: ***aufgewachsen* in einer kleinen provinz im sueden deutschlands, und unzufrieden mit ihrem einfachen und langweiligem leben**, entscheidet sie sich nach berlin zu ziehen. (*raised in a small province in the south of Germany and dissatisfied with their simple and boring life . . .*)

(5)5.m.S3.1027.b.cha: line 13.

*T: **friedman, *ein weltbekannter journalist von der new york times zeitung***, stellt den leser die phaenomen von globalisation und ihre auswirkungen vor. (*friedman, a world famous journalist at the new york times paper . . .*)

(6)7.DasElixierderNacht.cha: line 36.

*T: **nach der lektüre des zweiten (von drei) bänden, ebenso schillernd und fesselnd *geschrieben***, fürchte ich leider, daß diese trilogie den normalen fantasyleser nicht so stark ansprechen dürfte, wohl aber die an geschichte interessierten. (*after reading the second (of three) volumes, written in an equally luminous and captivating fashion . . .*)

(7)32.m.s3.1113.b.cha: line 38.

*T: **hamlet, *der immer zu viel denkt***, macht sich sorgen, dass dieser geist nicht sein vater sondern der tofel war. (*hamlet, who always reflects too much . . .*)

Lexical complexity in theme that results from expansion and manipulation of the noun phrase, a typical characteristic of academic register (Martin 1989; Schleppegrell 2004), is a major information-structuring device and is a sign of advanced writing capabilities. Analysis of the lexically complex themes across different levels demonstrates (Table 8.4) a steady increase in their use, a tripling at Level 5 compared with Level 3. A 40% increase of lexically complex themes in NS book reviews compared to the texts at Level 5 is also significant.

Table 8.4 Lexically complex themes across levels

	NS	Level 5	Level 4	Level 3
Lexically complex theme	9.9%	5.9%	3.5%	2%

Examination of the structural variety of nominal modifications that contributes to the complexity of lexical theme reveals the following development in the availability of modifying resources across levels.

While Level 3 writers complicate their themes only by means of relative clauses and prepositional phrases, Level 4 and Level 5 writers display a much richer array of modification resources. Apart from relative clauses and prepositional phrases, they use pre-nominal participial constructions, appositions and extended attributes. In fact, the only new type of noun modification observed in the NS data that is not encountered in NNS texts is the post-nominal participial construction presented as Example (6) above.

In book reviews, lexically complex themes like clausal themes are employed by writers as a strategy that addresses the two major communicative goals

Table 8.5 Structural variety in noun modification

NS	Level 5	Level 4	Level 3
– Relative clauses – Prepositional phrases and case modifications – Pre-nominal participials and adjectives – Appositions – Extended attributes – Post-nominal participials	– Relative clauses – Prepositional phrases and case modifications – Pre-nominal participials and adjectives – Appositions – Extended attributes	– Relative clauses – Prepositional phrases and case modifications – Pre-nominal participials and adjectives – Appositions – Extended attributes	– Relative clauses – Prepositional phrases and case modifications

of the genre: the necessity to present the content of the book in a succinct but logically clear way and the necessity to evaluate its content. How lexically complex themes function with regard to the first objective is illustrated by the example of the Content move, where lexically complex themes help address the challenge of revealing to the reader the right amount of detail about the plot and the protagonists of the book.

Lexically complex themes in the Content Move

Lexically complex themes figure prominently in the Content move, where they are used to mention important details about the book, happenings in the story and the protagonists' character as a point of departure for further predication or evaluation. Lexically rich themes contain a large amount of information that constitutes background in the sense that it is not presented as newsworthy or the focus of the content presentation. Despite the fact that these lexically complex 'background' elements occupy the place where given information is usually placed as a point of departure, they are often new information to the reader. They are, however, presentable in the theme because they are tied to a given noun head, most often in the form of pre- or post-nominal modifications of a noun phrase. In Prince's terms, this information can be categorized as *containing inferrables* (Prince 1981: 237). The following examples demonstrate how containing inferrables as part of a lexically complex theme helps present the content in a condensed, but at the same time logical way, without leaving out important and interesting details. In the following excerpts, the complex lexical theme provides additional information about the book (5.m.S3.9097.b.cha, ns4A.b.cha) and the characters, from introducing their names (ns5.b.cha) to evaluating their actions or personalities (7.DasElixierderNacht.cha, ns4B.b.cha).

5.m.s3.9097.b.cha; line 15. CONTENT

*T: **in dem in 1995 erschienen tatsachenroman 'die zwillinge'** handelt es um zwillingsschwestern die kurz vor dem zweiten welt krieg getrennt wurden. (*in the factual novel 'the twins', which appeared in 1995 . . .*)

ns4.b.cha: line. CONTENT

A*T:**sven regeners erstlingsroman herr lehmann, der im jahr 2001 im eichborn verlag publiziert wurde**, stellt in herzerfrischendem erzaehlstil das leben des frank lehmann zur zeit des mauerfalls in berlin vor. (*sven regner's first novel herr lehmann, published in 2001 by eichborn . . .*)

B*T:**zufrieden mit sich und seiner arbeit** strebt er nicht nach einer karriere und weltveraenderung, sondern nimmt mit wachem blick fuer das alltaegliche sich und seine umgebung war – und auf die schippe. (*content with himself and his work . . .*)

ns5.b.cha: line. CONTENT

*T: **eine junge frau, besagte veronika des buchtitels**, beschließt in diesem roman in der tat zu sterben. (*a young woman, the veronika of the title of the book . . .*)

7.DasElixirderNacht.cha: line. CONTENT

*T: **in die hände dieser skrupellosen männer der kirche** gerät, eher versehentlich, ein englischer apothekergehilfe, der durch zufall in den besitz eines geheimnisvollen pulvers gekommen ist, das tote zum leben erwecken kann (*into the hands of these unscrupulous men of the church . . .*)

Not only does the lexically complex theme provide the reader with additional significant details about the book or its plot, making the whole book review more informative; more importantly, it constructs a framework for interpretation of the rheme in the same sentence. This is very much in line with Haiman's proposal (1978) that topics, or in our terminology themes, present information as given at the time of utterance, so that they are 'givens by agreement' (cited in Schiffrin 1992: 162 from Haiman 1978: 584). As is evident from the following example, it is precisely the modification elements constituting the complexity of the theme and presented as given, even though they are not derivable from the previous discourse, that motivate the information in the rheme, in this case the actions of the protagonist.

ns10.b.cha
Content

*T: **der erzähler, der am anfang des romans von einem auto angefahren wird und sterbend auf der straße liegt**, nutzt <u>die letzten minuten seines lebens</u>, die geschichte seines eigenen lebens, das stark von der teilnahme an den studentenprotesten geprägt ist, mit den erzählunge anderer lebensgeschichten zu verweben. (*the narrator who at the beginning of the novel was hit by a car and lies dying in the street uses the last minutes of his life*)

*T: **der erzähler, der seinen lebensunterhalt als begräbnisredner verdient**, wird auch von aschenberger, der den damaligen idealen im laufe seines lebens, im gegensatz zu vielen seiner ehemaligen genossen, nicht abschwört, zum <u>leichenredner</u> bestellt. (*the narrator who makes his living as a funeral orator . . .*)

The idea that the narrator uses the last minutes of his life (the underlined rheme) to tell his story is based on the fact that the narrator is hit by the car and lies dying in the street, which is first introduced by means of a lexically complex theme. In the following T-Unit, information about the narrator being commissioned to speak at the funeral (the underlined theme) is motivated because the lexically complex theme first establishes the narrator as a funeral speaker. Use of lexically complex themes enables writers to manipulate or shape important information in a semantically hierarchical way by foregrounding some aspects and backgrounding others. In other words, lexically complex themes allow writers to include rich details but to do so in a way that does not obscure the overall structural pattern of the move. In this fashion their writing appears as structurally transparent or in Halliday's words 'crystalline' (1994: 224).

Conclusion and implications

Although complex themes are not the only factor in successfully organizing written discourse, analysis demonstrates that they are powerful instruments for creating coherent and cohesive texts. Manipulation of grammatical and lexical complexity of themes helps to present information in a way that displays certain organizational patterns associated with a particular communicative function of the move. In the Evaluation and Motivation moves, grammatically complex themes were shown to facilitate the construction of argumentative steps and enable evaluation, thus contributing to persuasion as the global coherence goal of the moves. In the Content move, packing information into lexically complex themes permits inclusion of interesting details about the plot in a fashion that avoids stringing them out across several clauses, which might result in loss of communicative focus. Instead, lexically complex themes organize texts consistently around ideational elements that function as communicatively crucial points of departure in a particular move; in turn these can be further elaborated or developed in rhemes. At the same time, employing longer themes is a cohesion strategy. Through thematization of subordinate clauses and extended noun phrases writers can connect previous discourse with the following, thereby producing a logically tighter line of reasoning.

Use of complex themes reveals an ability to strategically plan presentation of information that is indispensable for successful realization of communicative goals of written discourse in the secondary discourse contexts. Hypothesized to be a sign of advanced writing, a correlation between an increase in the use of complex themes and higher levels of language acquisition was confirmed by the study. This finding has implications for foreign language curricula that could include explicit instruction on the role of theme in structuring secondary discourse genres. When contextualized with regard to the role of coherence- and cohesion-building structures in achieving the communicative purposes of particular genres, such instruction can provide advanced foreign language learners with a convincing rationale for the use of 'more complex syntax' and more extended noun phrases and ultimately push them towards expanding their proficiency to include more complex registers of a foreign language.

Notes

1 Six of 54 book reviews were written on fiction books and thus included a narration about the events of the book. For information/issues books, the content of the book was presented in the Comment move.

2 Both native and non-native users of German tend to observe otherwise normative use of the German umlaut and ß quite variably in electronic contexts. No changes were made in these data, all the more so as orthographic accuracy was not in focus. Also, analysis programs typically reduce German capitalizations.

3 Because the examples focus on the nature of the generic moves these writers incorporate into their book reviews, translations do not reflect grammatical inaccuracies or other infelicities of expression in the original German.

References

Byrnes, H. and Sprang, K. (2004) 'Fostering advanced literacy: a genre-based, cognitive approach', in H. Byrnes and H. Maxim (eds), *Advanced Foreign Language Learning: A Challenge to College Programs*. Boston, MA: Heinle Thomson, pp. 47–85.

Carrell, P. (1982) 'Cohesion is not coherence'. *TESOL Quarterly*, 16, 479–88.

Christie, F. (1985) 'Language and schooling', in S. Tchudi (ed.), *Language, Schooling and Society*. Upper Montclair, NJ: Boynton/Cook, pp. 21–40.

Christie, F. (1986) 'Writing in schools: Generic structures as ways of meaning', in B. Couture (ed.), *Functional Approaches to Writing. Research Perspectives*. London: Pinter, pp. 221–39.

Connor, U. (1984) 'A study of cohesion and coherence in English as a second language students' writing'. *Papers in Linguistics: International Journal of Human Communication*, 17, 301–16.

Daneš, F. (1974) 'Functional sentence perspective and the organization of the text', in F. Daneš (ed.), *Papers on Functional Sentence Perspective*. The Hague: Mouton de Gruyter, pp. 106–28.

Drury, H. (1991) 'The use of systemic linguistics to describe student summaries at university level', in E. Ventola (ed.), pp. 431–57.

Eggins, S. (1994) *An Introduction to Systemic-Functional Linguistics*. London: Pinter.

Fahnestock, J. (1983) 'Semantic and lexical coherence'. *College Composition and Communication*, 34, 400–416.

Francis, G. (1989) 'Thematic selection and distribution in written discourse'. *Word*, 40, 201–21.

Fries, P. (1994) 'On theme, rheme and discourse goals', in M. Coulthard (ed.), *Advances in Written Text Analysis*. London: Routledge, pp. 229–50.

Gee, J. (1998) 'What is literacy?' in V. Zamel and R. Spack (eds), *Negotiating Academic Literacies: Teaching and Learning Across Languages and Cultures*. Mahwah, NJ: Lawrence Erlbaum, pp. 51–9.

Ghadessy, M. (1995) 'The relationship between thematic development and structural elements in a text', in M. Ghadessy (ed.), *Thematic Development in English Texts*. London: Pinter, pp. 129–47.

Haiman, J. (1978) 'Conditionals are topics'. *Language*, 54, 564–89.

Haiman, J. (1986) 'Constraints on the form and meaning of the protasis', in E. Traugott, A. ter Meulen, R. Snitzer and C. Ferguson (eds), *On Conditionals*. Cambridge: Cambridge University Press, pp. 215–27.

Halliday, M. A. K. (1989) *Spoken and Written Language*. Oxford: Oxford University Press.

Halliday, M. A. K. (1991) 'Linguistic perspectives on literacy: A systemic-functional approach', in F. Christie (ed.), *Literacy in Social Processes*. Northern Territory University, Darwin: Centre for Studies of Language in Education, pp. 2–23.

Halliday, M. A. K. (1994) *An Introduction to Functional Grammar* (2nd edn). London: Edward Arnold.

Halliday, M. A. K. and Hasan, R. (1976) *Cohesion in English*. London: Longman.

Halliday, M. A. K. and Hasan, R. (1989) *Language, Context, and Text: Aspects of Language in a Social Semiotic Perspective*. Oxford: Oxford University Press.

Halliday, M. A. K. and Martin, J. (1993) *Writing Science: Literacy and Discursive Power*. London: Palmer Press.

Halliday, M. A. K. and Matthiessen, C. M. I. M. (2004) *An Introduction to Functional Grammar*. London: Edward Arnold.

Hasan, R. (1995) 'The conception of context in text', in P. Fries and M. Gregory (eds), *Discourse in Society: Systemic Functional Perspectives.* Norwood, NJ: Ablex, pp. 183–285.

Hasan, R. (1996) 'What's going on: A dynamic view of context in language', in C. Cloran, D. Butt and G. Williams (eds), *Ways of Saying: Ways of Meaning. Selected Papers of Ruqaiya Hasan.* London: Cassell, pp. 37–50.

Hyland, K. (2000) *Disciplinary Discourses: Social Interactions in Academic Writing.* New York: Longman.

Khalil, A. (1989) 'A study of cohesion and coherence in Arabic college students' writing'. *System*, 17, 359–71.

Kuo, C. (1995) 'Cohesion and coherence in academic writing: From lexical choice to organization'. *RELC Journal*, 26, 47–62.

Labov, W. (1972) 'The transformation of experience in narrative syntax', in W. Labov (ed.), *Language in the Inner City.* Philadelphia, PA: University of Pennsylvania Press, pp. 354–96.

Lautamatti, L. (1987) 'Observations on the development of the topic of simplified discourse', in U. Connor and R. Kaplan (eds), *Writing Across Languages: Analysis of L2 Texts.* Reading, MA: Addison-Wesley, pp. 87–114.

Lautamatti, L. (1990) 'Coherence in spoken and written discourse', in U. Connor, and A. Johns (eds), *Coherence in Writing. Research and Pedagogical Perspectives.* Alexandria, VA: TESOL, pp. 29–43.

Lotfipour-Saedi, K. and Rezai-Tajani, F. (1996) 'Exploration in thematization strategies and their discoursal values in English'. *Text*, 16, 225–49.

Martin, J. (1984) 'Language, register and genre', in F. Christie (ed.), *Children Writing: Reader.* Geelong, Australia: Deakin University Press, pp. 21–30.

Martin, J. (1985) 'Process and text: Two aspects of human semiosis', in J. Benson and S. Greaves (eds), *Systemic Perspectives on Discourse.* Norwood, NJ: Ablex, pp. 248–74.

Martin, J. (1989) *Factual Writing.* Oxford: Oxford University Press.

Martin, J. (1997) 'Analysing genre: functional parameters', in F. Christie and J. Martin (eds), *Genre and Institutions. Social Processes in the Workplace and School.* London: Cassell, pp. 3–39.

Martin, J. (1998) 'Linguistics and the consumer: The practice of theory'. *Linguistics and Education*, 9, 411–48.

Martin, J. (1999) 'Mentoring semogenesis: "genre-based" literacy pedagogy', in F. Christie (ed.), *Pedagogy and the Shaping of Consciousness.* London: Cassell, pp. 123–55.

Martin, J. and Rose, D. (2003) *Working with Discourse. Meaning Beyond the Clause.* London: Continuum.

Mauranen, A. (1996) 'Discourse competence – Evidence from thematic development in native and non-native texts', in E. Ventola and A. Mauranen (eds), pp. 195–230.

Nwogu, K. and Bloor, T. (1991) 'Thematic progression in professional and popular medical texts', in E. Ventola (ed.), pp. 369–85.

Prince, E. (1981) 'Toward a taxonomy of given-new information', in P. Cole (ed.), *Radical Pragmatics.* New York: Academic Press, pp. 223–55.

Schiffrin, D. (1992) 'Conditionals as topics in discourse'. *Linguistics*, 30, 165–97.

Schleppegrell, M. (2004) *The Language of Schooling.* Mahwah, NJ: Lawrence Erlbaum.

Schmid, M. (1999) *Translating the Elusive. Marked Word Order and Subjectivity in English-German Translation.* Amsterdam/Philadelphia: John Benjamins.

Schneider, M. and Connor, U. (1990) 'Analyzing topical structure in ESL essays. Not all topics are equal'. *Studies in Second Language Acquisition*, 12, 411–27.

Steiner, E. and Ramm, W. (1995) 'On theme as a grammatical notion for German'. *Functions of Language*, 2, 57–93.

Tierney, R. and Mosenthal, J. (1981) 'Cohesion and textual coherence'. *Research in the Teaching of English*, 17, 215–30.

Ventola, E. (ed.) (1991) *Functional and Systemic Linguistics: Approaches and Uses.* Berlin/New York: Mouton de Gruyter.

Ventola, E. and Mauranen, A. (eds) (1996) *Academic Writing: Intercultural and Textual Issues.* Amsterdam/Philadelphia: John Benjamins.

Winterowd, R. (1970) 'The grammar of coherence'. *College English*, 31, 828–35.

Witte, S. (1983) 'Topical structure and revision: An exploratory study'. *College Composition and Communication*, 34, 313–41.

Witte, S. and Faigley, L. (1981) 'Coherence, cohesion, and writing quality'. *College Composition and Communication*, 32, 189–204.

9 The dialogic construction of meaning in advanced L2 writing: Bakhtinian perspectives[1]

Susan Strauss, Parastou Feiz, Xuehua Xiang and Dessislava Ivanova

This chapter presents an alternative approach to writing pedagogy, with a particular focus on ESL freshman composition. The approach rests dually on Bakhtin's notion of dialogicality and on the concept of languaging with its central focus on the socially interactive construction of meaning (for application to speaking development in advanced L2 learning, see Swain, this volume). It contrasts with existing pedagogical paradigms in freshman composition that conceptualize rhetorical skill development as critical thinking skill development and consider both as requisite to 'good writing'. We provide details of one ESL freshman composition course designed and implemented using this approach, including syllabus excerpts, activity descriptions, writing prompt samples and extracts from professional texts used in the class and from student writings.

Freshman composition in L1 and L2: Pedagogical goals and objectives

Courses in freshman composition are essentially designed to socialize entering college students into post-secondary literacy practices, in other words, to prepare students to effectively address the variety of reading and writing tasks that they will encounter across academic disciplines as well as those outside the academy (Crowley 1998; Lindemann 1993, 1995; UCLA website 2005). Overall, freshman composition programs nationwide tend to target such issues as: rhetorical knowledge, voice and genre; the writing process (e.g., invention, drafting and revision); and critical thinking skills, i.e., reasoning ability and argument analysis in the contexts of both reading and writing (e.g., Berlin 1987; Bizzell 1992; Ennis 1985; Johnson 1992; Pennsylvania State University English Department website 2005, henceforth PSU website; UCLA website 2005).

However, while such programs are inherently intended as a socializing medium, their stated goals and objectives imply a number of assumptions concerning best practices in composition pedagogy as well as precisely who is socialized through these programs and how (Atkinson and Ramanathan

1995; Schleppegrell 2004). These assumptions have implications for both L1 and L2 environments.

The first assumption is a theoretical and pedagogical one, i.e., that mastery of specific skills and abilities leads to the development of critical thinking and hence to 'good writing' (e.g., Lindemann 1993, 1995; PSU website; Schleppegrell 2004; UCLA website).

The second assumption is a cultural one: freshman writing programs are implicitly aimed at incoming students with native or native-like competence, who have already been socialized from early childhood into what Scollon and Scollon (1981) term mainstream 'essayist literacy' (see also, Atkinson and Ramanathan 1993; Farr 1993; Schleppegrell 2004), i.e., the discourse style that underlies the bulk of formal writing instruction in the United States. For students with mainstream literacy backgrounds, freshman composition serves to reinforce cultural literacy values and refine discourse skills that are already an integral part of their academic culture (Atkinson and Ramanathan 1995).

On the other hand, for students whose native language is not English or who speak a non-standard variety of English, freshman writing programs pose a more complex set of challenges (Atkinson and Ramanathan 1995; Schleppegrell 2004). International university students, for example, not only face this inherent culture-literacy gap, but research has also uncovered differences between L1 and L2 writers in multiple facets of their writing practices. Such differences have been noted in areas of metacognitive activities, e.g., overall task orientation, goal-setting and planning, as well as in areas affecting actual written products, e.g., topic introduction, organization, paraphrasing techniques, sentence structure, cohesion, modification (see Connor 1984; Leki 1992; Matsuda 1998; Silva 1993).

Writing classes specially designed for ESL freshmen acknowledge such differences while maintaining the same general goals and targeting the same skills- and abilities-based objectives as L1 composition. Thus, freshman composition syllabi for ESL writers contain many of the same types of assignments as the mainstream courses.

However, L2 writing courses do differ in terms of 'accommodation'. That is, in contrast with L1 instruction which emphasizes subtlety, inductive reasoning, complexity and style, L2 instruction tends to be more explicit, deductive and corrective, both in in-class activities and in instructor-written feedback. Class activities incorporate drills and practice for sentence-level mechanics, e.g., grammar, spelling, punctuation. Assignments are generally fewer in number and reduced in both scope and cognitive challenge. And the overall set of expectations with regard to student writing and cognitive development tends to be more relaxed. Atkinson and Ramanathan (1995) report a similar disparity in L1 and L2 writing programs administered at a large U.S. university.

Thus, L2 freshman composition echoes the predominant assumption underlying composition programs in general, i.e., that there is a direct relationship between the mastery of rhetorical skills, the development of critical thinking and the ultimate development of good writing. Nevertheless, a second assumption attenuates that echo. Since international students lack the linguistic, literacy and cultural experience in the host country, they are at a disadvantage; L2 instructors, therefore, should strive to achieve similar overall outcomes, while simplifying the materials and relaxing their academic standards.

What typifies writing pedagogy in general, then, is an ostensible disjunction between so-called critical thinking skills on the one hand, and the fundamental role of language on the other. Development of critical thinking skills is framed as an ability to simultaneously evaluate, analyse, put forth an argument and defend it logically and persuasively, all while conforming to the conventions of written English. The incongruity here is this: what is emphasized and explicitly labelled as desired 'skills' in composition programs is discussed and evaluated as cognitive abilities, detached from socially and contextually situated language use. In L2 environments, the incongruity is all the more conspicuous and potentially impedes growth and learning in the very areas that these programs target (for extensive discussion of the critical language base of thinking, see Matthiessen, this volume; for the link between language-based thinking and writing development, see Colombi, Ryshina-Pankova and Schleppegrell, this volume).

Furthermore, the learning processes typified in this approach to composition (both L1 and L2) centre predominantly, if not exclusively, on the individual learner engaged in the accomplishment of solitary activities (reading, writing, thinking, etc.). Students read and interpret texts, identify lines of argumentation and produce their own texts, with a view to ultimately mastering rhetorical skills *qua* cognitive skills, through continued practice and peer/teacher feedback on their output.

A dialogically discursive and literacy-centred approach to freshman writing

The pedagogical approach to freshman writing proposed in this chapter differs from traditional views in its underlying philosophy of learning. Learning, in this approach, is viewed as a socially achieved activity in which participants with varying degrees of expertise, experience and knowledge collaboratively and intersubjectively engage in meaning-making activities within and through dialogue (e.g., Claxton 2002; Kramsch 2000; Resnick *et al.* 1991; Swain and Lapkin 1995; Vygotsky 1986; Wells 1981, 1999, 2002; Wertsch 1985, 1991; for a similar position, see also Swain, this volume). It differs specifically in the primacy that it places on language; language not only as produced in written text or argument or analysis, but language as central to the mediation of cognition and thought (Vygotsky 1986; Swain, this volume). Thus, 'critical thinking' is reconceptualized here as the active and interactive activity of problem-solving, sense-making, questioning and experimentation, with and through language (for the link between learning a language and learning *through* and *about* a language, see especially

Matthiessen, this volume). Language is at once the medium through which external activity is internalized and the medium through which internal psychological activity is externalized.

The dialogic foundation of the approach is grounded in the Bakhtinian (1981, 1986) notion of 'dialogism', i.e., the belief that linguistic practices are dually shaped both by prior discourse and by the inherent addressivity of discourse. That is, any word or any utterance is a response to a preceding or potential word or utterance; it is inextricably connected and related to a complex network of other words or other utterances; words and utterances are addressive, in that they are directed to a recipient, intended or otherwise (see also Kramsch 2000: 139; Wells 1999: 104; and Wertsch 1991).

In fact, all discourse is dialogic (Bakhtin 1981, 1986).

> The speaker is not Adam, and therefore the subject of his speech itself inevitably becomes the arena where his opinions meet those of his partners . . . or other viewpoints, worldviews, trends, theories . . . The utterance is addressed not only to its own object but also to others' speech about it . . . [A]n utterance is a link in the chain of speech communication and it cannot be broken off from the preceding links that determine it both from within and without, giving rise within it to unmediated response reactions and dialogic reverberations. (Bakhtin 1986: 94)

Wertsch (this volume) provides a finer-grained perspective of Bakhtin's notion of dialogue by drawing a distinction between 'local dialogue' and 'generalized collective dialogue'. For Wertsch, local dialogue involves the face-to-face or co-present intermingling of one speaker's words and/or utterances with words or utterances of another, similar in concept to the common interpretation of the term 'dialogue'. Generalized collective dialogue, by contrast, refers to the process(es) by which individual and collective voices of others not immediately present in an interactional exchange are reflected in the current discourse. By positing such a distinction, Wertsch makes it clear in the context of language instruction (L1 and L2) that knowledge of grammar, sentence structure and lexis (local dialogue) represents merely a fraction of the issue. Advanced language learning requires mastery and a true understanding of the dynamic, heterogeneous conflux of the various 'speech genres' that belong to the generalized collective dialogue. The type of dialogic socialization that takes place within this alternative pedagogical framework, then, shifts away from the perspectives of learning 'genre', 'logical reasoning' and 'usage conventions' as static, decontextualized and rigidly homogeneous notions, toward one in which language use is simultaneously dynamic, context-shaping and meaning-driven.

From the point of view of participant-orientation, the locus of attention in this paradigm shifts from the domain of the individual (i.e., text, reader and writer), characteristic of the traditional paradigm, to the socially interactive and co-participatory learning environment in which issues and opinions are presented, negotiated, transformed and re-analysed. Moreover, the textual focus shifts from a form-centred, rhetorical analysis of overt and covert lines of argumentation within given readings to both local and global analyses of discourse and text. This engages learners in a diverse array of semiotic elements (at macro and micro levels) involved in the construction of discourse and in the creation of meanings within conventionally appropriate modes and genres. In this way, 'the act of learning [becomes a] fundamentally functional and usage-based social practice rather than . . . a primarily analytical and rule-based individual activity' (Byrnes and Sprang 2004: 49).

The literacy-centredness of the course design mirrors much of what Kern (2004: 7) proposes for a literacy-based curriculum, including, for example: the incorporation of a wide range of texts (spoken, written, visual and audiovisual); an integrated focus on linguistic, cognitive and social dimensions of language use; the goal for students to recursively analyse, interpret and transform discourse; and the provision of structured guidance in the appropriate active interpretation and production of contextually situated spoken and written discourse. Kern's perspective on literacy erases the boundaries between language skills and content, because it is language use and the discursive creation of meaning and text that becomes the 'object of analysis and reflection' (2004: 7).

This view of literacy is especially appropriate to the design of ESL freshman composition courses given the central role of discourse in general; the integration of all language modalities (i.e., not focusing solely on reading and writing); the focus on texts as semiotic tools for meaning-making; the inclusion of diverse media as such 'texts'; and carefully considered explicit pedagogical guidance in the analysis, interpretation and production of discourse.

Languaging in action – one ESL freshman composition course

In the remainder of this chapter, we provide results of a research project in which this dialogic, literacy-centred approach to writing was implemented at a large university in the Eastern United States. The course was instituted by the authors between 2001 and 2004 as a research-based experimental approach to second language writing pedagogy within an otherwise traditional ESL writing program. A precursor to the course was designed by the first author and implemented for a period of approximately three years as a credit-based ESL freshman composition course at a major West Coast university. The course is a credit-based ESL version of the required freshman composition course. It is offered through the applied linguistics department of that university.

The results reported here concern two sections of this alternative writing course from the spring of 2004. A total of 22 students were enrolled in the sections, each taught by a different instructor. The instructors were applied linguistics graduate students who had been trained in the approach and involved with the project since its inception in 2001. The course name and number remained identical to that used for the traditional ESL composition courses, though the assignments and pedagogical approaches differed. Approximately 14 of the 22 students (67%) had just arrived in the United States the previous semester; the remainder had spent a number of years in the country, some having attended high school here. The class met twice weekly throughout the 15-week semester, with each class period totalling 75 minutes. The course required a total of four multiple-draft essays. The essay assignments were progressively sequenced in terms of level of difficulty, cognitive and linguistic challenge and genre, from the first assignment, a personal narrative, to the final one, an argumentative synthesis:

- Assignment #1: personal narrative (2-draft minimum)
- Assignment #2: comparison and contrast (3-draft minimum)
- Assignment #3: critique (3-draft minimum)
- Assignment #4: argumentative synthesis (3-draft minimum)

Other required writing activities included informal journal entries, typically related to various drafts of the writing assignments, and a final portfolio cover piece that accompanies the complete portfolio of student writing at the end of the semester.

In the sections that follow, we provide examples of how dialogic meaning construction is achieved through carefully selected readings and other semiotic media, through class discussions, as well as through assignments that challenge students to make connections between texts, their personal experiences and their own writing – all while maintaining a keen sensitivity to the concept of the 'reader'.

Literacy and dialogue: readings, films, student writings

As a means of illustrating the centrality of literacy and dialogue, we focus here on the first two essay assignments, i.e., the personal narrative and the comparison and contrast essay. The same required texts were critical for accomplishing both tasks: 'House Calls', by Lewis Thomas (1995); 'Labyrinthine', by Bernard Cooper (1997); and 'Happiness', by Czeslaw Milosz (2001). The texts had been selected for the quality and style of writing, the various perspectives the authors conveyed about their childhoods and the effect those perspectives had on the now adult writers.

Accompanying each reading was a vocabulary list that reflected specialized or metaphorical uses of otherwise more common terms or terms that were crucial to the narrative line and/or imagery of each essay. The order of appearance of the entries reflects the order of appearance in the original piece; where necessary, the entry contained both a paraphrased dictionary definition in addition to commentary pertaining to its significance in the writing, as in Figure 9.1, from 'Labyrinthine'. 'Labyrinthine' depicts family, ageing and the passage of time. It evokes multiple curvilinear and maze-like images, emerging first as playthings on surfaces and in fabric and later as wrinkles in the faces of loved ones.

labyrinth:	a maze, a complicated network of paths and obstacles, designed as a puzzle to go successfully from the starting point to the end point. Labyrinth and maze mean essentially the same thing; labyrinth is a more sophisticated term (followed by sample maze from 'Tesseract's Mazes' entitled 'Sunface').
paisley:	a pattern of curved figures – paisley was a popular design in the U.S. in the 1960s and 1970s for clothing and furniture. It is still seen as a design on ladies' scarves (followed by two sample photos). . .
redwood:	tall tree with reddish wood; these trees live for a very long time, and their age is determined by counting the number of growth rings in their trunk (followed by a photo of a tree in the forest and one of a cross-section of trunk showing rings).
a late child:	a child born to parents who are already older, e.g., in their late 30s or early 40s.

Figure 9.1 'Labyrinthine'. Vocabulary list (excerpt)

Following the reading and a general discussion of its content, students engaged in an in-class activity involving clips from three films: 'To Kill a Mockingbird', 'A River Runs Through It' and 'By Hook or By Crook'. Video clips are incorporated into the coursework as additional text because of their interconnected semiotics of themes, images, music, lighting and voice. Such complex symbol systems stimulate 'associations, stored memory, meanings, and expressions' that transcend what might be touched off through linguistic means alone (Salomon 1994: 118).

Further, the clips served to complexify and enrich the literacy repertoire with which students work. That is, the activity involved a task in which students were asked to discover features that all six pieces had in common, not just the three readings. Students generated lists of specific elements shared by all six texts; they were then guided to collaboratively pinpoint one global thematic commonality. They ultimately discovered that all the texts recounted memories of childhood through the voices of a now-adult narrator, but predominantly through the eyes of that narrator as a child. That is, the images created in all six pieces clearly reflect perspectives, thoughts and lived childhood experiences as filtered through the mature perceptual lens of an adult, voiced through the language of an accomplished writer, and depicted through diverse visual means in the films.

This class activity also involved a task in which students were guided by the instructors to discover constitutive elements of narrative, essentially: tellability (e.g., Kramsch 1989; Polanyi 1979); sequential (and/or logical) progression of events (e.g., Labov 1972; Ochs 1994); and type/level of detail relevant to the story, e.g., granularity and density of descriptions concerning key protagonists, places, events, landscapes and so forth (Talmy 1995).

The next class session opened with instructors asking students which of the three readings they related to most deeply and why; small groups were organized on that basis for students to discuss the reasons for their choices; and finally, a full class discussion ensued in which students expressed their reactions, opinions and personal histories. Mediated by the instructor, the group created a space for mutual trust, patience, openness, respect and compassion. Students realized at a very early stage in the course that within this space it was both safe and constructive to risk exposing personal feelings, views and experiences. This quality of openness and mutual respect undergirds the entire 'community of writers' approach discussed in Strauss (in preparation). Such a sense of trust and respect is crucial for the class to cohere early on and to establish the collaborative workshop atmosphere central to this approach to writing.

The discussion also served as an embodied demonstration of the links that exist between text (written and film) and experience, feeling memory and imagination. That is, students witnessed first-hand how they and their classmates extracted meaning from texts, applied it to aspects of their own lives and came to evaluate their experiences newly from a fresh perspective, tinged by texts, images or their classmates' reactions.

Both dialogicality and languaging are clearly central not only to the activity itself or to the entire foundation of the class, but ultimately, and most importantly, to the students' cognitive development. Students' utterances and opinions and viewpoints are all links in the *chain of speech* (Bakhtin 1986: 94); and it is through these types of problem-solving activities of languaging that their thinking is 'articulated and transformed into artifactual form' (Swain, this volume).

Students were then ready to locate with more precision those passages and images in the three essays that caused them to be so moved. This involved recapitulating the general impression created by each and uncovering the overall message and purpose in the writings. It also involved analysis of specific micro-level instances of language use that collectively mesh into the literary whole of each writing. Students were asked to 'mine' (Greene 1992) each text for patterns in structure, paragraph development, imagery and metaphor and repetition.

How such 'mining' might be accomplished can be seen from a marked-up copy of the entire text of Bernard Cooper's 'Labyrinthine' in Appendix A. Of the three assigned essays, the language used in 'Labyrinthine' was the most metaphorically rich and the most elaborated in terms of imagery, granularity of detail, coherence, sentence structure, paragraph development and organization. Its extensive linguistic representations of temporality and its use of images viewed from the child's as well as the adult-writer's perspective, all contribute to the tellability of the story, create its coherence and underscore the reason why Cooper wrote it to begin with.

The annotations in Appendix A reflect three main themes: 1) literal and symbolic reference to the 'mazes' throughout Cooper's life, 2) the progression of time, and 3) the concept of 'inevitability' signalled once as a counterfactual in the third paragraph: 'If only I'd known a word like "inevitable", since that's how it felt to finally slip into the innermost room', and which re-surfaced in the final paragraph.

To aid students in such 'language mining', instructors provided preliminary guidance in the search for such patterns of imagery, perspective, repetition and temporal continuity/temporal shifts. Thereafter, students marked their texts using a colour-coding system to highlight the various linguistic and rhetorical patterns that they noticed upon re-reading. Students were urged to use this discourse–analytic strategy in approaching texts for the remaining readings of the class and in other reading they might engage in.

The key notion underscored in this and related activities is that of choice. Students experienced first hand the creative power of lexical, semantic and syntactic choice in the creation of a cohesive and engaging piece of writing. They came to realize and understand the power of language, and more importantly, to relate to this power in a language other than their native one. As a result, they discovered that the entire essay was built on a single metaphor, the metaphor of a maze. The maze represented at once Cooper's boyhood passion, his parents' diminishing health and mental lucidity, and the turns and traps and puzzles that his own life holds for him as he comes to terms with the 'inevitability' of his own ageing process.

Assignment 1 : Personal narrative

After engaging in similar discourse analytic activities with the remaining two essays, students received the prompt for writing assignment 1. The text of the prompt is reproduced in (1) below:

(1) Prompt – essay #1 – personal narrative

> Write a personal narrative in which you describe a childhood memory (or set of memories) which deeply influenced some aspect of your thinking or feeling as an adult. The memory could be one of an event, a person, something someone said to you, etc., which shaped part of who you are today. Be sure to include a description of the setting/background; provide as much detail as possible (or necessary) to make this a coherent, well organized, and engaging piece of writing. (Length: approx. 3 pages, double-spaced).

The assignment requires students to produce a narrative with a particular purpose, similar in nature to the three samples they had read in preparation for the task. Students were engaged in the task and many produced essays with perspectives, voices and images that echoed what they had read in the three original pieces. There was no explicit instruction to emulate any of the discursive techniques used in the readings. However, Kiamin designed his entire piece around a single metaphor: models. For Kiamin, models represented his childhood hobby, the relationship that he had had with his brother, the conflicts that existed between them and the lessons that life has since taught him. An abridged version of Kiamin's essay appears in (2) below. The sections in bold represent discernable influence from Cooper's original piece, including his use of temporal adverbials, sentence structure and the progression in perceptions and feelings from those of an innocent child to those of a young man.

(2) Sample essay #1 – personal narrative. Author: Kiamin (excerpt)

Models

Whenever I had spare time, I would usually pay a visit to the model-making shop with my younger brother, though it is a long way from my home. The shop is not so large, around 600 square feet, I suppose. **Because I was small, boxes of models built up like mini-buildings to me . . .**

I did not know why I was so crazy with those tiny and brittle plastic models that time. I could even recognize which were the new arrivals . . . Robot models were my favorites. They had attractive body. Their shields look impermeable. **Energy swords, laser guns, and long-range rifles were kinds of weapons. All these were boys' favorites**.

Every time I opened the package of a model, there would always be scissors and model gel next to me. The gel had a strong pungent chemical smell. However, I showed no rejection to it because I often needed its help.

. . . As I was getting older, I started to make models other than robots, like replicas of vehicles, helicopters and military hardwares. **My brother was a color-painting expert. He gave lives to those monotonous plastic models only with the help of different colors of markers, model dyes, and brush tools**. I took a deep breath and then had a great sense of satisfaction every time I finished . . . a model . . .

When I was young, conflicts were the common things that always existed between me and my younger brother. My parents always stood on my brother's side. They always wanted to me to make the concession. The result was: **My models became the real sufferers. Nearly every time when I had an argument with my brother, he usually showed off his power by destroying my models like a giant monster destroying a city. My models often had broken arms or legs after every war**. He never had any punishment . . . Nevertheless, I was generally gaining more and more patience. Maybe this was only training. **My brother initially wanted to train me to tolerate unequal things. I thought this assumption could make me feel better. As I am growing up, I realize that tolerance is really important in maintaining and enhancing human relationships**.

It is a bitter sadness and loneliness with the models in my home right now. I am in the States and my brother is in Canada. Both of us have gone and those models are left in our home far away from where we are. It is hard for them to see any arguments between me and my brother again in the near future.

Kiamin's essay is a transparent representation of the type of 'local' and 'generalized collective' dialogicality (Wertsch, this volume) that the course strives for. It also reflects Kiamin's willingness to take risks in writing. He used language creatively, mirroring much of what he had just recently read and reacted to himself. In particular, he assimilated the notion that symbols from life's experiences create a network of meaning that links simple, concrete images and form to abstract thought and feelings – all of which change and transform over time. Further, Kiamin's essay attempts to move his readers much in the same way that he had been moved by language and image and structure.

The next section provides samples of another type of discursive dialogicality that emerged through face-to-face interactions among peers and teacher as well as through the creative process of writing.

Assignment 2: Comparison and contrast

The comparison and contrast essay was designed to challenge students to think beyond surface-level observations and characterizations and to make connections in readings, films and overall life experiences that they might not otherwise have made or even noticed; in other words, to advance their own understanding of text and life experiences reflected therein through language and dialogue, ultimately arriving at new discoveries and renewed understandings. The assignment builds on previous knowledge and pushes students to think more deeply about a piece of writing or a film and to 'see' much more in that piece than what meets the eye. An abridged version of the prompt appears in (3):

(3) Prompt – essay #2 – comparison and contrast (abridged)

> So far, we have read three essays ('Labyrinthine', 'House Calls' and 'Happiness') and viewed three introductory film clips ('To Kill a Mockingbird', 'A River Runs Through It' and 'By Hook or By Crook'), in which a narrator has described *an important realization or change in thinking* that has occurred to him or her through time. In each case, memories are described through the eyes of the narrator as a child, and *these memories are all relevant to an important ideological shift – a shift in philosophy, political views, social views, personal views,* etc. For this essay, compare and contrast ***the type of ideological shift*** across **at least two works** (though you may use **up to four**) – any combination of essays/films is fine.

The readings all centre on some important realization or shift in thinking on the part of the narrator; none mentions this explicitly. However, the shifts are relevant to each author in very different ways. This assignment urges students to determine what has changed in each narrator's life and why, and to ascertain as well the relevance of any shifts to his/her current perspectives on life, profession, nature, etc. It requires meaningful and thoughtful synthesis of texts.

Assignment 2 had a minimum requirement of three drafts. At the first draft, the writing tends to stray from the assignment expectations; it is often an experimental stage where students grapple with logic and ideas in a rather unfocused way. Second drafts reflect the greatest amount of substantive change as students orient themselves to the task (through classroom-based activities) and build a foundation on which to develop their ideas. By the third draft, the most conspicuous revisions affect introductions and conclusions as well as the overall organization of ideas. At this level, which follows both peer review and individual teacher–student conferences, students present more concrete examples, elaborate their points in more depth and support general statements with compelling illustrations.

In the case of the first draft of Assignment 2, a majority of students attempted to respond to the prompt and locate the 'ideological shift', but they were generally unsuccessful. An example, written by Yoon, appears in (4):

(4) *Comparison/Contrast. Author: Yoon, draft #1 (excerpt)*

> What makes his father successful? It's the **skepticism** of his father to value of medicine, surgery, and even doctor's ability to heal patient . . . In his essay, Lewis Thomas basically gives the significant meaning to his father's inclination which is represented by **skepticism**; also, he appraises his father's **skepticism** as the obvious reason for successful reputation. Throughout the whole essay, he re-illuminates his father's skepticism by relating to his own experiences or by introducing others' words . . .

Essentially, Yoon wrote an analytic summary. He provided no evidence of having discovered any type of 'shift'.

Typically, the instructor read through all first drafts without marking any comments on them but synthesizing in her or his mind suggestions for improving them; these suggestions were presented at the next class meeting. The remainder of that period was dedicated to collaborative peer review that took the following form: instructors determined the best first draft of the class, anonymized it and distributed copies to each class member for a collaborative analysis of what made it 'good' as a first draft, in other words, those aspects of the writing that address the task effectively. Once the analysis has been exhausted, discussion moves to aspects that are less effective, concluding in collaborative suggestions (mostly by student peers) for further development and improvement at the next draft level. Students infer what revisions might be appropriate for their own next drafts, strictly on the basis of this in-class activity and without written feedback or separate student conferences.

Excerpt (5) is from the exemplary essay (i.e., the 'best' first draft) used for the collaborative peer review. It was the only essay of the two sections that addressed the assigned task:

(5) *Comparison/Contrast. Author: Leonardo, draft #1 (excerpt)*

> . . . I noticed a conceptual resemblance in the way how the change in the author's view of 'Labyrinthine' goes from enthusiasm to **skepticism**, and how closely this exemplifies in the viewpoint of the author's father in 'House Calls' who people thought of his work as miracle maker but he **kept skeptic**, and *cleverly modest* against all the 'mazes' he could solve as a doctor . . . Is (it) a matter of '***prudence***' the author of Labyrinthine would say, as he describes how now, middle aged, he understands the lines in his parents' faces, as a way of saying 'these are all the mazes I've been through, I'm done'? The doctor would just add, it is a matter of ***prudence*** not to

help someone, it is a matter of ***prudence*** not to try again and again, as if each day was a new one . . .'

Leonardo pinpointed shifts in Cooper's ('Labyrinthine') and Thomas' ('House Calls') thinking by voicing the internal dialogues of each writer. Choosing original words and images from one and interweaving them into the other (*skepticism* = from 'House Calls' and *maze* = from 'Labyrinthine') was his strategy for establishing a foundation for his comparison and contrast. Leonardo also invented altogether new attributes that had not figured explicitly in either original text, e.g., *prudence*, as a potential common thread linking Cooper's and Thomas' dispositions, and the descriptor for Thomas' father as *cleverly modest*. In so doing his language use exhibits an impressively complex thought process as well as a facility to create new meanings for himself and for his readers.

After circulating and discussing Leonardo's draft during the collaborative peer review, all students revised their own essays. On the whole, students' second drafts in both sections displayed marked improvement. One telling example is Yoon's second draft, an excerpt of which appears in (6). By now Yoon has started to gain linguistic and conceptual control over what he had originally meant by the term 'skepticism' (Thomas' original word choice) and has succeeded in teasing apart the various nuances of this term as they might apply to each writer's psychological state and/or shift. The draft is now on task with respect to the overall writing assignment.

(6) Comparison/Contrast. Author: Yoon, draft #2 (excerpt)

'. . . In 'House Calls', the main ideological shift is made as the author **allowed new values** to what his father had done during his childhood. He could become to re-evaluate his father's **skeptical philosophy** to the medicine . . . The readers repeatedly see his father's ***prudence*** on the effectiveness of medical therapy . . . The author first recognized his father's those kinds of doubts to the medicine as **skepticism**, but later he came to **realization** that it was **honesty** to himself as a doctor and ***modesty** to his ability*. And he **evaluated** his father's carefulness to the medicine as a basis of building a reputation . . . Through his childhood, the author saw his father's innate reluctance to the medical treatment somewhat in **value-free point of view** and **simply dealt with skepticism. Yet later he became to understand and evaluated it while the author got through his own medical experiences and heard about his father's accomplishment from people who knew his father** . . .'

Yoon's increasingly deeper understanding of 'skepticism' seems to have been mediated by Leonardo's word choice of 'prudence' and 'modest(y)', among other factors. Yoon has appropriated these terms, but applies them in his own way, from his own perspective and using his own voice.

The foregoing progression of excerpts reveals the extent to which the combined processes of dialogue and languaging mediate learning and advance understanding and cognitive development. In the context of one L2 freshman composition course, we have observed that, through a carefully crafted variety of discursive activities and controlled guidance through those activities, students are indeed being socialized into the literacy practices of American English. The so-called 'critical thinking skills' characteristic of 'essayist literacy' are replaced by immersion in and mastery of the literacy conventions of various genres, which are 'at heart, ways of thinking, reasoning, analyzing, and problem solving' (Byrnes, 2005 p.c.).

Conclusion and implications

Through the foregoing discussion, we hope to have provided a new perspective on pedagogical approaches to freshman composition. The approach presented here is founded on the pivotal importance of dialogue and languaging as crucial to the socially interactive construction of meaning. We have also illustrated that the emergence of such socially semiotic dialogue is largely, if not entirely, dependent upon an open academic environment which encourages the constructive exchange of personal understandings, interpretations, opinions and experiences among participants.

The examples and textual excerpts illustrate precisely how dialogue with text engenders new dialogues and the creation of new texts – a cognitive outcome not easily achieved through solitary and isolated acts of reading, writing and responding to specific prompts.

With respect to L2 writers, the approach may be all the more powerful in that the socialization process into the rhetorical practices of the academy is accomplished as a natural outcome of the collaborative activity. We find that, as advanced L2 writers, students can be expected to both recognize specific discursive and semiotic patterns in oral and written text and to actively and creatively appropriate those patterns in their own discourse, choosing them with confidence and control. Within these semiotically grounded activities of reading, analysing, discussing, writing and revising, students are guided to make 'meaning-driven choices that occur within conventions of use', rather than simply being 'creative in a sociolinguistically non-recognizable fashion' (Byrnes and Sprang 2004: 51). The notion of 'choice' is key both in students' interpretive analysis and in their own language use.

Notes

1 The authors are deeply indebted to Heidi Byrnes for her invaluable comments on previous versions of the manuscript. This chapter would not have taken the shape that it now has without Heidi's keen theoretical and editorial insights and her very patient reading of our earlier drafts.

References

Atkinson, D. and Ramanathan, V. (1995) 'Cultures of writing: an ethnographic comparison of L1 and L2 university writing/language programs'. *TESOL Quarterly*, 29, 539–68.

Bakhtin, M. M. (1981) *The Dialogic Imagination: Four Essays by M. M. Bakhtin.* Austin,

TX: University of Texas Press. (Edited by M. Holquist; translated by C. Emerson and M. Holquist.)

Bakhtin, M. M. (1986) *Speech Genres and Other Late Essays*. Austin, TX: University of Texas Press. (Edited by Caryl Emerson and Michael Holquist; translated by Vern W. McGee.)

Berlin, J. (1987) *Rhetoric and Reality: Writing Instruction in American Colleges, 1950–1985*. Carbondale, IL: Southern Illinois University Press.

Bizzell, P. (1992) *Academic Discourse and Critical Consciousness*. Pittsburgh, PA: University of Pittsburgh Press.

Byrnes, H. and Maxim, H. H. (eds) (2004) *Advanced Foreign Language Learning: A Challenge to College Programs*. Boston: Heinle Thomson

Byrnes, H. and Sprang, K. A. (2004) 'Fostering advanced L2 literacy: a genre-based cognitive approach', in H. Byrnes and H. H. Maxim (eds), pp. 47–85.

Claxton, G. (2002) 'Education for the learning age: a sociocultural approach to learning to learn', in G. Wells and G. Claxton (eds), pp. 21–33.

Connor, U. (1984) 'A study of cohesion and coherence in English as a second language students' writing'. *Papers in Linguistics*, 17, 301–16.

Cooper, B. (1997) 'Labyrinthine', in I. Frazier and R. Atwan (eds) *The Best American Essays – 1997*. NY: Houghton Mifflin, pp. 33–6.

Crowley, S. (1998) *Composition in the University: Historical and Polemical Essays*. Pittsburgh, PA: University of Pittsburgh Press.

Ennis, R. (1985) 'Critical thinking and the curriculum'. *National Forum*, 65, 28–31.

Farr, M. (1993) 'Essayist literacy and other verbal performances'. *Written Communication*, 10, 4–38.

Greene, S. (1992) 'Mining texts in reading to write'. *Journal of Advanced Composition*, 12, 151–70.

Johnson, R. (1992) 'Critical reasoning and informal logic', in R. Talaska (ed.) *Critical Reasoning in Contemporary Culture*. Albany, NY: State University of NY Press, pp. 69–88.

Kern, R. G. (2004) 'Literacy and advanced foreign language learning: rethinking the curriculum', in H. Byrnes and H. H. Maxim (eds), pp. 2–18.

Kramsch, C. (1989) 'Discourse and text: a narrative view of the foreign language lesson', in J. E. Alatis (ed.) *Language Teaching, Testing, and Technology: Lessons from the Past with a View Toward the Future (GURT '89)*. Washington, DC: Georgetown University Press, pp. 69–80.

Kramsch, C. (2000) 'Social discursive constructions of self in L2 learning', in J. Lantolf (ed.), *Sociocultural Theory and Second Language Learning*. Oxford: Oxford University Press, pp. 133–53.

Labov, W. (1972) 'The transformation of experience in narrative syntax', in *Language in the Inner City*. Philadelphia: University of Pennsylvania Press, pp. 354–96.

Leki, I. (1992) *Understanding ESL Writers: A Guide for Teachers*. Portsmouth, NH: Heinemann.

Lindemann, E. (1993) 'Freshman composition: no place for literature'. *College English*, 55, 311–16.

Lindemann, E. (1995) 'Three views of English 101'. *College English*, 57, 287–302.

Matsuda, P. K. (1998) 'Situating ESL writing in a cross-disciplinary context'. *Written Communication*, 15, 99–121.

Milosz, C. (2001) 'Happiness', in C. Milosz, *To Begin Where I Am. Selected Essays*. New York: Farrar, Straus and Giroux, pp. 20–6. (Edited and with an introduction by Bogdana Carpenter and Madeline G. Levine.)

Ochs, E. (1994) 'Stories that step into the future', in D. Biber and E. Finegan (eds) *Sociolinguistic Perspectives on Register*. New York: Oxford University Press, pp. 106–35.

Pennsylvania State University, Department of English website (2005). *http://English.-la.psu.edu/undergraduate/compjump.htm*. Accessed 28 December 2005.

Polanyi, L. (1979) 'So, what's the point?' *Semiotica*, 25, 208–41.

Resnick, L., Levine, J. and Teasley, S. (eds) (1991) *Perspectives on Socially Shared Cognition*. Washington, DC: American Psychological Association.

Salomon, G. (1994) *Interaction of Media, Cognition, and Learning: An Exploration of How Symbolic Forms Cultivate Mental Skills and Affect Knowledge Acquisition*. Hillsdale, NJ: Lawrence Erlbaum.

Schleppegrell, M. J. (2004) *The Language of Schooling: A Functional Linguistics Perspective*. Mahwah, NJ: Lawrence Erlbaum.

Scollon, R. and Scollon, S. (1981) *Narrative, Literacy, and Face in Interethnic Communication*. Norwood, NJ: Ablex.

Silva, T. (1993) 'Toward an understanding of the distinct nature of L2 writing: the ESL research and its implications'. *TESOL Quarterly*, 27, 657–77.

Strauss, S. (in preparation) Creating a community of second language writers: An advanced level curriculum design for freshman composition in second language contexts.

Swain, M. and Lapkin, S. (1995) 'Problems in output and the cognitive processes they generate: a step towards second language learning'. *Applied Linguistics*, 16, 371–91.

Talmy, L. (1995) 'Narrative structure in a cognitive framework', in J. F. Duchan, G. A. Bruder and L. E. Hewitt (eds), *Deixis in Narrative: A Cognitive Science Perspective*. Hillsdale, NJ: Lawrence Erlbaum, pp. 421–60.

Thomas, L. (1995) 'House Calls', in L. Thomas. *The Youngest Science: Notes of a Medicine Watcher*. New York: Penguin Books, pp. 12–18.

UCLA Website (2005) Student Academic Services Analytical Writing Placement Examination. *www.ucop.edu/sas/awpe/requirement.html*. Accessed on 28 December 2005.

Vygotsky, L. (1986) *Thought and Language*. Cambridge, MA: MIT Press.

Wells, G. (1981) *Learning through Interaction: The Study of Language Development*. Cambridge: Cambridge University Press.

Wells, G. (1999) *Dialogic Inquiry: Toward a Sociocultural Practice and Theory of Education*. Cambridge: Cambridge University Press.

Wells, G. (2002) 'Inquiry as an orientation for learning, teaching and teacher education', in G. Wells and G. Claxton (eds), pp. 197–210.

Wells, G. and Claxton, G. (eds) (2002) *Learning for Life in the 21st Century. Sociocultural Perspectives on the Future of Education*. Malden, MA: Blackwell.

Wertsch, J. (1985) *Vygotsky and the Social Formation of Mind*. Cambridge, MA: Harvard University Press.

Wertsch, J. (1991) 'A sociocultural approach to socially shared cognition', in L. B. Resnick, J. M. Levine and S. D. Teasley (eds), pp. 85–100.

Appendix

BERNARD COOPER

Labyrinthine

FROM THE PARIS REVIEW

→ WHEN I DISCOVERED my first maze among the pages of a coloring book, I dutifully guided the mouse in the margins toward his wedge of cheese at the center. I dragged my crayon through narrow alleys and around corners, backing out of dead ends, trying this direction instead of that. Often I had to stop and rethink my strategy, squinting until some unobstructed path became clear and I could start to move the crayon again.

I kept my sights on the small chamber in the middle of the page and knew that being lost would not be in vain; wrong turns only improved my chances, showed me that one true path toward my reward. Even when trapped in the hallways of the maze, I felt an embracing safety, as if I'd been zipped in a sleeping bag.

Reaching the cheese had about it a triumph and finality I'd never experienced after coloring a picture or connecting the dots. If only I'd known a word like "inevitable," since that's how it felt to finally slip into the innermost room. I gripped the crayon, savored the place.

The lines on the next maze in the coloring book curved and rippled like waves on water. The object of this maze was to lead a hungry dog to his bone. Mouse to cheese, dog to bone — the premise quickly ceased to matter. It was the tricky, halting travel I was after, forging a passage, finding my way.

→ Later that day, as I walked through our living room, a maze revealed itself to me in the mahogany coffee table. I sat on the floor, fingered the wood grain, and found a winding avenue through it. The fabric of my parents' blanket was a pattern of climbing ivy and, from one end of the bed to the other, I traced the air between the tendrils. Soon I didn't need to use a finger, mapping my path by sight. I moved through the veins of the marble heart, through the space between the paisleys on my mother's blouse. At the age of seven I changed forever, like the faithful who see Christ on the side of a barn or peering up from a corn tortilla. Everywhere I looked, a labyrinth meandered.

→ Soon the mazes in the coloring books, in the comic-strip section of the Sunday paper, or on the placemats of coffee shops that served "children's meals" became too easy. And so I began to make my own. I drew them on the cardboard rectangles that my father's dress shirts were folded around when they came back from the cleaner's. My frugal mother, hoarder of jelly jars and rubber bands, had saved a stack of them. She was happy to put the cardboard to use, if a bit mystified by my new obsession.

The best method was to start from the center and work outward with a sharpened pencil, creating layers of complication. I left a few gaps in every line, and after I'd gotten a feel for the architecture of the whole, I'd close off openings, reinforce walls, a slave sealing the pharaoh's tomb. My blind alleys were especially treacherous; I constructed them so that, by the time one realized he'd gotten stuck, turning back would be an exquisite ordeal.

My hobby required a twofold concentration: carefully planning a maze while allowing myself the fresh pleasure of moving through it. Alone in my bedroom, sitting at my desk, I sometimes spent the better part of an afternoon on a single maze. I worked with the patience of a redwood growing rings. Drawing myself into corners, erasing a wall if all else failed, I fooled and baffled and freed myself.

→ Eventually I used shelf paper, tearing off larger and larger sheets to accommodate my burgeoning ambition. Once I brought a huge maze to my mother, who was drinking a cup of coffee in the kitchen. It wafted behind me like an ostentatious cape. I draped it over the table and challenged her to try it. She hadn't looked at it for more than a second before she refused. "You've got to be kidding," she said, blotting her lips with a paper napkin. "I'm lost enough as it is." When my father returned from work that night, he hefted his briefcase into the closet, his hat wet and drooping from the rain. "Later," he said (his code word for "never") when I waved the banner of my labyrinth before him.

It was inconceivable to me that someone wouldn't want to enter a maze, wouldn't lapse into the trance it required, wouldn't sacrifice the time to find a solution. But mazes had a strange effect on my parents: they took one look at those tangled paths and seemed to wilt.

I was a late child, a "big surprise" as my mother liked to say; by the time I'd turned seven, my parents were trying to cut a swath through the forest of middle age. Their mortgage ballooned. The plumbing rusted. Old friends grew sick or moved away. The creases in their skin deepened, so complex a network of lines, my mazes paled by comparison. Father's hair receded, Mother's grayed. "When you've lived as long as we have . . . ," they'd say, which meant no surprises loomed in their future; it was repetition from here on out. The endless succession of burdens and concerns was enough to make anyone forgetful. Eggs were boiled until they turned brown, sprinklers left on till the lawn grew soggy, keys and glasses and watches misplaced. When I asked my parents about their past, they cocked their heads, stared into the distance, and often couldn't recall the details.

Thirty years later, I understand my parents' refusal. Why would anyone choose to get mired in a maze when the days encase us, loopy and confusing? Remembered events merge together or fade away. Places and dates grow dubious, a jumble of guesswork and speculation. *What's-his-name* and *thingamajig* replace the bright particular. Recollecting the past becomes as unreliable as forecasting the future; you consult yourself with a certain trepidation and take your answer with a grain of salt. The friends you turn to for confirmation are just as muddled; they furrow their brows and look at you blankly. Of course, once in a while you find the tiny, pungent details poised on your tongue like caviar. But more often than not, you settle for sloppy approximations — "I was visiting Texas or Colorado, in 1971 or '72" — and the anecdote rambles on regardless. When the face of a friend from childhood suddenly comes back to me, it's sad to think that if a certain synapse hadn't fired just then, I may never have recalled that friend again. Sometimes I'm not sure if I've overheard a story in conversation, read it in a book, or if I'm the person to whom it happened; whose adventures, besides my own, are wedged in my memory? Then there are the things I've dreamed and mistaken as fact. When you've lived as

36 *Labyrinthine*

long as I have, uncertainty is virtually indistinguishable from the truth, which as far as I know is never naked, but always wearing some disguise.

Mother, Father: I'm growing middle-aged, lost in the folds and bones of my body. It gets harder to remember the days when you were here. I suppose it was inevitable that, gazing down at this piece of paper, I'd feel your weary expressions on my face. What have things been like since you've been gone? Labyrinthine. The very sound of that word sums it up — as slippery as thought, as perplexing as the truth, as long and convoluted as a life.

10 Learning advanced French through SFL: Learning SFL in French

Alice Caffarel

For the past fifteen years or so a growing number of researchers and teachers have begun to use systemic-functional (SF) descriptions of languages other than English (LOTE) to teach languages and linguistics at the tertiary level. Indeed, this volume is testimony to that effort (see particularly the contributions by Colombi, Crane, Ryshina-Pankova, Schleppegrell and Teruya). Even as these scholars have access to a considerable body of work on English teaching and teaching about English from a SF perspective (e.g., Christie 1983, 1990; Christie and Rothery 1979; Christie and Unsworth 2005; Halliday 1979; Hasan and Martin 1989; Kress 1982; Martin 1985, 1986; Martin and Painter 1986; Martin and Rose 2005; Melrose 1995), little work is available on how the teaching of and through SF descriptions of LOTE might inform and facilitate development in a non-native language that is not English.

In this paper, I want to investigate at some depth how SFL has been applied to advancing the learning of the French language in the Department of French Studies at the University of Sydney, Australia, particularly with intermediate to advanced learners of the language. The following interrelated issues are of particular concern: (1) how learning about the French grammatical system in French as a meaning-making resource might enhance students' ability to gain sophisticated levels of awareness about the multilayers of meaning in texts of various types; (2) how increased awareness of text-oriented understandings of meaning–form relationships might affect their sense of choice in meaning-making, an issue that is of particular importance at the advanced level; (3) how awareness of the meaning-making capacities in the L2 environment might enhance their overall semiotic capacities, including, of course, their L1 capacities; and (4) how this kind of heightened awareness might be a contributing factor to continued L2 development.

I begin by exploring in a general way the suitability of SFL as a pedagogical tool for teaching and learning a second language. I then locate the place of linguistics in the Department of French Studies at the University of Sydney and, by implication, the possibility of an SFL approach within such a departmental context. I conclude by focusing on one particular course,

'*Introduction à la linguistique*', examining its course structure, syllabus, pedagogies and learning outcomes potential, relating these points to more general opportunities and challenges that arise from teaching SFL in French.

Learning French through SFL

As a theory of language as meaning potential, SF theory is a powerful pedagogical tool for teaching and learning a second language, particularly at the advanced level. With its explicit focus on text rather than sentence-level grammatical features, it is centrally concerned with meaning-making in particular contexts of use. That means that, in contrast with the grammatical systems that customarily underlie L2 pedagogy, SFL right from the beginning considers meaning and form as inseparable and sees them as instantiated in texts. Furthermore, because it links language to context, both situational and cultural, SFL seems particularly well suited to make advanced language learners aware of the kinds of choices that exist at various strata of the language system and the contexts of situation and culture. As Matthiessen (this volume), following Halliday, explicates, it does so particularly transparently through its theoretical apparatus of metafunctions and register, as well as the instantiation and stratification dimensions. The multiple dimensions of SF theory provide the second language learner with different pathways for exploring the construal and contruction of meaning in texts and for expanding his/her meaning potential in the second language. As Halliday points out:

> In all language education, the learner has to build up a resource. It is a resource of a particular kind: a resource for creating meaning. I call it a 'meaning potential'. Whether someone is learning the mother tongue, learning to read and write, learning a second or foreign language, learning the language of science or mathematics, or learning the styles of written composition – all these are forms of meaning potential. What the learner has to do is to construe (that is, construct in the mind) a linguistic *system*. That is what is meant by 'language as system': it is language as stored up energy. It is a language, or some specific aspect of a language, like the language of science, in the form of a *potential*, a resource that you draw on in reading and writing and speaking and listening – and a resource that you use for learning with. How do you construe this potential, and how do you use it when you've got it? You build it up, and you act it out, in the form of *text*. 'Text' refers to all the *instances* of language that you listen to and read, and that you produce yourself in speaking and in writing. (1999: 7) (Emphases in original)

It is that emphasis on meaning potential that makes SFL a powerful framework for teaching the grammar of particular languages and applying it to the analysis of the meanings of texts. This strong orientation towards meaning also motivated the choice of the linguistics courses offered in the Department of French at the University of Sydney and, more particularly, the opportunity to use SFL as an instructional framework.

Linguistics in French studies

The Department of French Studies at the University of Sydney offers a variety of streams, which allows students to choose not just literature and language courses but also courses in linguistics, social sciences and francophone studies. I am responsible for the linguistics stream and have been able to use SFL in four different courses: *'Introduction à la linguistique'*, *'La linguistique textuelle'*, *'Deconstructing French texts'* and *'La grammaire du texte'*. The first course introduces students to the linguistic system as a stratified model of language in context with a particular focus on grammar. The second course focuses on the textual systems of cohesion. Both courses are offered to second-year advanced students and third-year intermediate students. 'Deconstructing French texts' is offered only to third-year advanced students and explores the construal of ideology in news media. This course combines a SFL approach with French approaches to discourse analysis. 'Grammaire du texte' is an honours seminar which attracts both literature and linguistics students and explores how grammatical patterns in literary texts construe a particular literary style.

All four courses are text-based; students explore how lexicogrammatical and/or semantic choices vary according to use in different text types and how different lexicogrammatical patternings create second-order meanings in a text, e.g., meanings of a social, ideological, philosophical or literary order. These courses aim not only to provide students with linguistically based conceptual tools for critical or literary analysis; deep down, they also aspire to facilitate students' continued language development through the expansion of their meaning potential in French. The contribution of learning a discipline in a particular language in order to enhance learning that language is often covert in language departments. By and large, language development is seen as independent from the learning of literature, linguistics and the social sciences; indeed, these disciplines are in many language departments taught in English in pursuit of more enrolments. While linguistics and literature courses in the Department of French Studies are referred to as 'content' courses and are, therefore, presented as distinct from the 'language' courses, the Department is committed to teaching these courses in French. From my perspective, the 'content' courses are not distinct from, but an extension of, the 'language' courses: their function is to bring students to advanced and specialized levels of French in different domains.

Inherently, any course about language presents language as both 'the medium of learning' and 'the substance of what is being learnt' (see Halliday 1999: 2), even though programs may typically privilege one to the detriment of the other – thereby tearing apart the underlying relationship of realization between the two. The pervasive dichotomy between 'content' and 'language' found in language departments is not surprising; indeed it echoes positions taken in many formal theories of language, such that semantics and grammar are to be treated in isolation. For example, Lyons (1968: 135) argued that 'nothing but advantage can come from the methodological separation of semantics and grammar'.

By contrast, in this paper I will argue the opposite: that everything can be gained from making explicit the natural relationship that exists between meaning and grammar, and in particular in the context of second language learning. In addition, the teaching of linguistics as the study of language as a system of choice highlights to students the relationship between language as a system and language as a text instance, thus enhancing their overall understanding of language production while building up their French language potential.

One of the objectives of the linguistic courses offered in the Department of French Studies is precisely to teach how choices at the level of semantics are realized in the grammar and instantiated in different text types. In other words, students learn to make choices in accordance with what they can mean in different situation types. They learn to 'think grammatically' (Halliday 2002: 370) and talk linguistically.

Learning about language through interaction and reflection

The linguistic courses offered in the Department of French Studies are all conducted as two-hour workshops over a 13-week semester. This is to facilitate interaction between students themselves and between the students and myself so that they learn to take on different roles in the classroom and reflect on their learning practices. Thus classroom activity is metafunctionally motivated and foregrounds language as a means for interacting (interpersonal metafunction) and for making sense of what students learn, that is, for interpreting experience (experiential metafunction). For that reason students are encouraged to talk about language among themselves, to question and challenge the lecturer, to reconstruct the reality encoded in texts in various ways and to interpret the meanings of various text types. Over the years, students' insights into the analysis of texts have contributed to my own research on French grammar and built up new resources for me and the students.

As Rowland notes:

> . . . where the knowledge that research produces is seen and is offered to students, as being tentative, open to reinterpretation or containing insights that can be applied more widely, the ways that students relate to this knowledge are potentially significant to the lecturer's own research.
>
> In other words, the approach to teaching that is linked most closely to research is one in which significance is given to what the students have to say, and opportunity is provided for their voices to be heard. (2000: 24)

The intertwinement of teaching/learning language and researching language in class is a powerful means of challenging students' knowledge and generating interaction. I will demonstrate how that manifests itself in a particular classroom setting by using the French systemic grammar course offered at the University of Sydney, '*Introduction à la linguistique: Approche fonctionnelle*', as the place of inquiry.

Learning about French, learning through French, learning French

Matthiessen (this volume) reiterates the three aspects of learning language discussed in Halliday (1999) as involving not only learning language but learning through language and learning about language.

> 'Learning language' means, of course, learning one's first language, plus any second or foreign languages that are part of the curriculum: including both spoken and written language – initial literacy, composition skills and so on. Here, language is itself the *substance* of what is being learnt. 'Learning through language' means using language, again both spoken and written, *as instrument*: as the primary resource for learning other things – language across the curriculum, in other words. 'Learning about language' means studying language as an *object* in order to understand how it works: studying grammar, semantic, phonetics and so on. Here language is a domain or branch of knowledge . . . (Halliday 1999: 21, emphases in original).

The course '*Introduction à la linguistique: approche fonctionnelle*' has learning about French as its primary focus: learning French then becomes a concurrent outcome of learning about French through the medium of French. One of the first and most important insights students gain about language in this course is that it is indissociable from context. As Halliday (1999: 22) points out, 'In all educational learning, learners are being required to predict both ways: to predict the text from the context, and to predict the context from the text.'

Indeed, SF theory makes explicit the indissociability of language from its context by positing context as a higher semiotic plane which is realized by language. For example, Figure 10.1 introduces students to the different strata of language and the realizational relationship existing between strata, the metafunctional diversification of content (semantics and lexicogrammar) into interpersonal, experiential and textual functions, and the situational variables of field (*champ*), tenor (*teneur*) and mode that provide the contextual information for making predictions about the linguistic features of a text.

Explicating that relation between context and language, Halliday states:

> [The field, tenor and mode of discourse] are the general concepts needed for describing what is linguistically significant in the context of situation. They include the subject-matter, as an aspect of the 'field of discourse' – of the whole setting of relevant actions and events within which the language is functioning – for this is where subject-matter belongs. We do not, in fact, first decide what we want to say, independently of the setting, and then dress it up in a garb that is appropriate to it in the context, as some writers on language and language events seem to assume. The 'content' is part of the total planning that takes place. There is no clear line

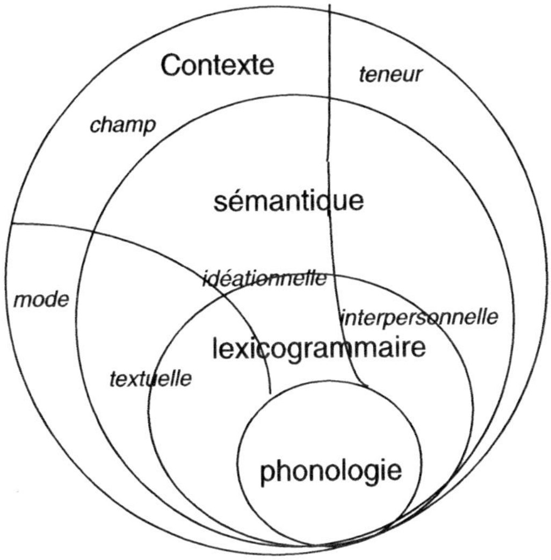

Figure 10.1 *La relation entre langage et contexte: réalisation*

> between the 'what' and the 'how'; all language is language-in-use, in a context of situation, and all of it relates to the situation, in the abstract sense in which I am using the term here. (Halliday 1978: 33)

Contrary to the generally held notion that classrooms are impoverished environments and that they are, therefore, rightly contrasted with the 'real world', this particular classroom is made into an excellent point of departure for illustrating the relationship between the situational variables of field, tenor and mode and the linguistic choices within the experiential, interpersonal and textual metafunctions: (1) how the 'institutional setting', i.e., the teaching of linguistics at the university level, influences choices in transitivity and lexis (experiential meaning); (2) how the 'relationship between participants', i.e., the variation in status and knowledge between students and teacher, influences the dynamics of the exchange in the classroom and the choices in mood (interpersonal meaning); and (3) how 'the channel of communication adopted and role of language', here spoken instruction, influences textual choices and the organization of the flow of information in the classroom (textual meaning). The following is an example of the kind of text the teacher might offer in the early part of the semester:

Text 1

Un aspect important de la théorie SF est la relation naturelle qui existe entre la sémantique et la lexicogrammaire. Les choix lexicogrammaticaux servent à réaliser des choix sémantiques. La relation qui existe entre les deux est parfois congruente, parfois métaphorique. C'est le fait que grammaire et sémantique ne sont pas toujours alignés qui rend la langue un outil de communication flexible et puissant.

Toute énonciation est en partie prévisible à partir du contexte, du Champ du discours (l'activité dont on parle), de la Teneur du discours (les relations qui existent entre interlocuteurs) et du Mode du discours (le rôle du langage/type de langage)

[An important aspect of SF theory is the natural relationship that exists between semantics and lexicogrammar. Lexicogrammatical choices serve to realize semantic choices. The relationship that exists between the two is sometimes congruent and sometimes metaphorical. It is the fact the grammar and semantics are not always aligned that makes language a flexible and powerful communication tool.

All text instance is in part predictable from context, from the field of discourse (the activity taking place), the tenor of discourse (the nature of the relationship between participants), and the mode of discourse (the role/type of language).] (Author translation)

An analysis of the linguistic resources of the above text reveals how a classroom situation motivates particular choices in the grammar. Experiential (transitivity) functions are indicated in brackets with lexical items in bold, textual functions (Themes) are italicized and interpersonal (mood and modality) choices are unmarked: the text is monologic, with the speaker giving general information to students about the linguistic theory they will be studying. The mood of the text is 'declarative' and there is no expression of attitude.

The experiential systems of transitivity provide choices for construing different domains of experience: the material domain (doings and happenings), the mental domain (thinking and perceiving), the verbal domain (saying and telling) and the relational domain (being and having). The following text construes experience as relations between entities and, predictably, processes are of the relational type.

(1) [Value] *Un **aspect important** de la **théorie SF*** [Proc:relational:identifying] est [Token] la **relation naturelle** [[qui **existe** entre la **sémantique** et la **lexicogrammaire**]].
(2) [Token] *Les **choix lexicogrammaticaux*** [Proc:relational:identifying] **servent** à **réaliser** [Value] des **choix sémantiques**.
(3) [Carrier] *La **relation** [[qui **existe** entre les deux]]* [Proc:relational:attributive] est [Attribute] parfois **congruente**, parfois **métaphorique**.
(4) C'est [Attributor/Agent] *le fait [[que **grammaire** et **sémantique** ne sont pas toujours **alignés**]]* qui [proc: relational:attributive:effective] **rend** [Carrier] la **langue** [Attribute] un **outil** de **communication flexible** et **puissant**.
(5) [Carrier] *Toute **énonciation*** [Proc:relational:attributive] est [Attribute] en partie **prévisible** [Circumstance] à partir du **contexte**, du **Champ du discours** (**l'activité** dont on **parle**), de la **Teneur du discours** (les **relations** qui **existent** entre **interlocuteurs**) et du **Mode du discours** (le **rôle du langage/type de langage**).

This kind of lecture style presentation presents many characteristics of written language even though in the classroom context it is channelled orally: all clauses are clause simplexes with a high lexical density: 7.4 lexical items per clause on average (see Halliday 1985 on spoken and written language).

In response to the macro-functional question of 'What is happening here?' (field of discourse) one might answer that the lecturer is introducing the students to new concepts and ideas about language. She does so by using relational clauses (identifying and attributive) to define and describe new linguistic concepts. A high proportion of the lexis is predictably technical with many terms specific to linguistics. While the lecturer is talking, students are engaged in listening. But it is easy to predict that, if this kind of discourse were to continue without some form of activity in the form of questions, illustrations and analysis, students would rapidly switch off. One way of motivating students and attracting their interest in complex phenomena is to make them active by reflecting back on language as it is used in the classroom. By applying the new notions learnt in the text to the text in question, students are explicitly faced with how semantics and grammar are related and how contextual variables motivate linguistic choices in the language of the lecturer. Thus, the text produced by the lecturer becomes a pedagogical device for generating interaction about context and text and grammatical metaphor in a situation where, typically, there would be little dialogue.

The linguistic choices that can be predicted from a 'lecture type' classroom situation as it is found in Text 1 are summarized in the table below:

Table 10.1 Situation type and text type: instantiation dimension

Contextual variables	Linguistic features
Field: 'What is taking place' Teacher introducing students to theory and ideas about languages	Participants: linguistic concepts Transitivity: relational clauses – defining, naming, describing
Tenor: 'Who is taking part' The lecturer, 'linguistic expert', is giving information to students, 'non-expert in linguistics'	Mood: declarative Speech roles: – Lecturer: speaker/giver of information – Students: listener/demander of information
Mode: 'What role language is playing' The function of the text is to present and explain ideas about language in a French linguistics class.	The language of instruction is French, the medium is spoken and the language 'written', with a high lexical density and no clause complexes.
	The themes are mostly unmarked and the text tends to have a linear progression where the Rheme of the previous sentence becomes the Theme of the following sentence. This kind of thematic progression is typical of explanatory texts.

For students to become active participants, the classroom situation must regularly change. Indeed the above text, which is typical of the kind of texts produced in a lecture context, although informative, is not conducive to students' participation. The classroom situation itself must vary so that a variety of texts where theory and practice interact are produced by both students and teacher. Thus, students are able to take on different speech roles, such as telling, asking, explaining, challenging or interpreting. This is precisely how students are introduced to the interpersonal function of language: by being made aware of what kind of exchanges take place in the classroom and what they can do in order to become active participants/learners. Biggs (1999: 76) notes that 'learner activity and interacting with others are two characteristics of rich teaching/learning contexts'.

However, even though students are introduced to the MOOD systems of French and to notions like 'congruent' and 'metaphorical' (see Halliday 1994: Chapter 10) by reflecting on the interaction that goes on in the classroom, they are not explicitly taught in this particular course how to analyse the interpersonal structures of clauses in terms of Subject, Complement and Adjuncts (see Caffarel 2006 on the interpersonal structure of the French clause). Two factors account for that decision, restriction in time and the nature of the chosen texts: in the mostly monologic news items and narratives ideational and textual resources tend to have greater prominence. For that reason the interpersonal metafunction, as illustrated through interpersonal resources (i.e., MOOD and MODALITY options), is presented primarily in natural occurrences within the activities of the course itself, and less so in the texts analysed. In other words, the interpersonal metafunction germinates as students work collaboratively on the various analytical tasks, and this is where students build up their interpersonal resources and reflect on their use of interpersonal metaphors.

We will now have a closer look at the course syllabus.

Learning French systemic-functional grammar

Learning grammar as a meaning-making resource is not something that Australian students find natural. The majority have never studied grammar at school, and what little they know comes from learning French as a second language where they are essentially introduced to grammar as a list of rules for writing correctly. Some may have studied linguistics elsewhere but usually within a formal framework where grammar is again interpreted as a system of rules.

The initial workshop of the semester is organized around a discussion on what linguistics is and what students think it is, a general comparison between different approaches to the study of language depending on one's objectives. Students explore what kinds of grammatical analysis different theories provide and how much semantic information these analyses can reveal. This is followed by a general introduction to the SF model of language in context, with illustrations from the classroom discourse. The objectives for this course are stated as:

> *Initier les étudiants au langage en tant que système significatif à partir du modèle théorique systémique fonctionnel. Ce cours a pour but de donner aux étudiants des connaissances linguistiques explicites qui leur permettront de comprendre le processus de communication, ainsi que d'analyser et d'interpréter les significations des ressources lexico-grammaticales utilisées dans des textes de registres différents.*

[To initiate students to language as a system for meaning based on systemic functional theory. This course aims at providing students with explicit linguistic knowledge that will allow them to understand the process of communication as well as analyse and interpret the meanings of the lexicogrammatical resources used in texts of different registers.]

As the program below shows, the course is organized around the metafunctions, with a particular focus on the logical, experiential and textual metafunctions. Students are introduced to the different metafunctions (interpersonal, ideational and textual) in week 2. They are also introduced to the grammatical system as a meaning-making resource and the notion of rank so that grammatical systems can be located in relation to metafunction and rank. Before moving onto the logical metafunction and the clause complex system in week 3, students explore complexity at various ranks and learn to identify groups and phrases and group and phrase complexes with examples taken from texts such as:

> ||*Aujourd'hui,*| *le rallye Paris-Dakar*| *traverse*| *le désert.*|| *Mais ces voitures, ces motos, ces camions [[bariolés par les slogans publicitaires]]* | *croisent parfois* |*dans le même désert* |*des caravanes [[composées seulement d'hommes et de bêtes]].* ||*(Decaux 1987: 17)*
>
> Today, the rally Paris-Dakar crosses the desert. But these cars, motorbikes, trucks [[splashed with advertising]] pass sometimes in the same desert caravans [[composed only of men and animals]].[Author translation]

This text illustrates the use of embedding and word complexes as a means of packaging more information into the nominal group. Once students become familiar with dividing texts into clauses and clauses in groups and phrases, they move onto the logical, experiential and textual metafunctions, and conclude with an overview of the interpersonal metafunction and interpersonal metaphors. Inasmuch as this course structure is analogic to the rhetorical organization used in Caffarel (2006), the following commentary by Caffarel is instructive:

> The choice of logical resources as point of departure for this description of French grammar is motivated by its discourse orientation. In a sense, the rhetorical development of this book maps onto the analytic process: the first step in analysing the lexicogrammatical resources of a text consists in dividing that text into clauses before we can proceed to the metafunctional analysis of each clause. Thus, in a

Semaine 1:	Qu'est-ce que la linguistique? Les deux grandes familles linguistiques: formelle (ex. Chomsky) et fonctionnelle (ex. Halliday) Le signe linguistique (Saussure) Les différents niveaux du système linguistique: sémantique, grammaire, phonologie. Contexte et registre (champ, teneur et mode du discours)
Semaine 2:	A quoi sert le langage? Les fonctions du langage Qu'est-ce que la grammaire? Qu'est-ce qu'une phrase? Les constituants de la phrase Le groupe nominal
Semaine 3:	Les phrases complexes (1)
Semaine 4:	Vacances de Pâques
Semaine 5:	Les phrases complexes (2) Devoir 1 (à rendre en fin de semaine 6)
Semaine 6:	La fonction idéationnelle: les systèmes de TRANSITIVITÉ La transitivité nucléaire et la transitivité circonstancielle: le modèle transitif
Semaine 7:	Les types de procès (1): actifs, verbaux & mentaux, et relationnels
Semaine 8:	Les types de procès (2): Les types de circonstances Devoir 2 (à rendre en fin de semaine 10)
Semaine 9:	Le modèle ergatif: La notion d'agence. Qui est responsable? (Distribution du sujet du dernier devoir, à rendre en fin de semaine 14)
Semaine 10:	Les Thèmes
Semaine 11:	Thèmes marqués et non-marqués
Semaine 12:	La métaphore interpersonnelle
Semaine 13:	Analyses

Figure 10.2 Course syllabus: '*Introduction à la linguistique: approche fonctionnelle*'

> discourse-based account of grammar it seems natural to first explore the resources that serve to combine clauses into clause complexes in texts. Furthermore, the principles behind clause complexing give us a means of identifying the clause, that is, the highest ranking grammatical unit. In addition, logical resources are 'fractal' types in the lexicogrammatical system in that they may manifest in other ranks: we may combine words and groups into complexes in the same ways that we combine clauses. (2006: 5)

In conjunction with learning each metafunction, students are made aware of how the foregrounding of certain grammatical patterns construes second-order meanings, an awareness that becomes an important aspect of the discussion and that is also an integral part of the interpretation of all text types. Construal of a second-order meaning through the logical metafunction, for example, is found in Louis-René des Forêts' *Le Bavard* (1947), as discussed and further analysed in Caffarel:

> 'Un bavard' is someone who talks a lot, a chatterbox. The novel is composed of a succession of extensive clause complexes which foreground the dynamic and fluid nature of talk which is central to the theme of this text. However unlike in natural talk the internal structure of each clause in a complex is itself complex, with complex group structures and a refined (not ordinary) vocabulary, making the text highly literary despite a logical structure very similar to that of casual conversation, as shown by the two extracts presented below. (2006: 49–50)

From *Le Bavard* (Louis-René des Forêts, 1947, p. 8) (A translation of the entire passage is provided below.)

1a	α							*On me demandera peut-être*
1b	″β	α						*si j'ai entrepris de me confesser*
1c		×β	α					*pour éprouver cette sorte de plaisir un peu morbide*
1d			=β	1				*dont je parle*
1e				+2	α			*et que je comparerais volontiers à celui [[que recherchent quelques personnes raffinées]]*
1f					=β	α		*qui, avec une lenteur étudiée, caressent du bout de l'index une légère égratignure*
1g						=β	1	*qu'elles se sont faite sciemment à la lèvre inférieure*
1h							+2	*ou qui piquent de la pointe de la langue la pulpe d'un citron à peine mûr.*

They will maybe ask me || if I began to confess || to experience this kind of pleasure a bit morbid || that I am talking about || and that I would happily compare to the one [[sought by some refined people]] || who, with a carefully designed slowness, caress with the tip of the index finger a slight scratch || that they have knowingly done to their lower lip || or prick with the tip of the tongue the pulp of a lemon hardly ripe. [Author translation]

[This abstract] is particularly pertinent because the narrator 'talks' about his writing style.

From *Le Bavard* (Louis-René des Forêts, 1947, pp. 9–10) (A translation of the entire passage is provided below.)

1a	1							*Mon goût me porte naturellement vers le style allusif, coloré, passionné, sombre et dédaigneux*
1b	+2	α						*et j'ai pris aujourd'hui, non sans répugnance, la résolution de laisser de côté toute recherche formelle,*
1c		×β	α					*de sorte que je me trouve écrire avec un style*
1d			=β	1				*qui n'est pas le mien;*
1e				=2	α			*c'est à dire que j'ai écarté tous les charmes dérisoires*
1f					=β	α		*dont il m'arrive parfois de jouer,*
1g						+β	1	*tout en sachant bien [[ce qu'ils valent]]:*
1h							+2	*ils ne sont les fruits que d'une habileté assez ordinaire.*

My taste leads me naturally towards an allusive, coloured, passionate, sombre and contemptuous style || and I have taken today, not without repugnance, the resolution to leave aside all formal research, || so that I find myself writing with a style || which is not mine; || that is, I have brushed away all the derisory charms || with which I sometime play || knowing well [[what they are worth]]: they are only the fruits of quite an ordinary skill. [Author translation]

The symbols next to each clause (e.g., 1, 2, α, =, etc.) are used to mark clauses in terms of clause complexing. Arabic numerals (1, 2, 3 . . .) stand for parataxis, and Greek letters (α, β, γ . . .) for hypotaxis. The logico-semantic relations are symbolized as follows: elaboration by the = sign, extension by +, enhancement by ×, projection of idea by ' and projection of locution by " (see also Figure 10.3).

Discussion and analysis in French of French texts and of how lexicogrammatical choices make meanings in context certainly challenge the students' conception of grammar. At the same time it challenges their linguistic abilities, inasmuch as they are exposed to the French language and metalanguage as language of instruction and as course content. Students have found that having the courses in French (see comments in Appendix A) improves their communicative (both spoken and written) skills in French, while learning about French improves students' overall semiotic capacities and critical skills (see Appendix A). As in many university classes, students in this class come with different levels of ability in French; some speak fluently, others have only studied French for two or three years. While such ability differences might at first appear to be problematic they can, in part, be resolved when students are told that accuracy of expression will not be considered in their formal assessment. Once students know that their 'grammar' in the traditional sense will not affect their mark, but that it is their understanding of grammar as a meaning-making resource that will be tested, they feel more at ease in expressing themselves, and their French gradually improves. To conclude this overview of '*Introduction à la linguistique*' I will now look at some of the issues related to teaching SFL in French.

Teaching SFL in French

As previously indicated, the course on French systemic-functional grammar is conducted in French and students are given course notes in French; however, all their other readings are in English. To help them deal with this problem, students are given a bilingual glossary of SF terms with definitions in French; Table 10.2 taken from Caffarel (2006) provides an example.

Table 10.2 shows that, in many instances, translation of the metalanguage from English into French is straightforward. Yet, more complex situations do occur. For instance, the ergative function 'Medium' could perhaps be translated as '*Moyen*' but this term is already used to refer to a circumstance of

Table 10.2 A sample from the English–French glossary of SF terms

Clause. Fr: **Phrase**. Rang supérieur du stratum lexicogrammatical.

Clause complex. Fr: **Phrase complexe**. Combinaison de phrases reliées par parataxe ou hypotaxe et non par enchâssement; la combinaison des phrases se fait selon le mode d'organisation logique de la métafonction idéationnelle. Plusieurs phrases peuvent être, par exemple, reliées par coordination pour former une phrase complexe.

Lexicogrammar. Fr: **Lexicogrammaire**. Combinaison de la grammaire et du lexique (vocabulaire); l'ensemble des ressources pour signifier verbalement. En dehors de la linguistique systémique, la grammaire et le lexique sont presque toujours traités comme des modules distincts. La théorie systémique interprète le lexique comme la grammaire la plus fine (see 'delicate' below) ou spécifique.

MOOD. Fr: **Modes/types de phrase**. Région interpersonnelle (*interpersonal*); la grammaticalisation des fonctions élocutives (*speech functions*) dans la phrase.

Realisation. Fr: **Réalisation**. Relation interstratique (entre les strata).

Instantiation. Fr: **Actualisation**. Relation intrinsèque (à l'intérieur d'un seul et même stratum).

Stratum. Fr: **Stratum**. Un système ou un ordre d'abstraction particulier au language: sémantique, lexicogrammaire et phonologie sont les trois strata du système linguistique dans la théorie systémique de Halliday. Les strata sont reliés par réalisation (interstratique). Par exemple, le stratum de la sémantique est réalisé par la lexicogrammaire.

Means (*circonstance de Moyen*) and also to middle clauses (*phrases moyennes*). A term is found to be appropriate when it reflects the meaning/function of what it stands for and when it makes sense to the students. The Medium is defined as the function 'through which the process is actualized, and without which there would be no process at all' (Halliday 1994: 163). The term that students thought best reflected the function/meaning of the Medium was '*Véhicule*'; importantly, in that choice for the French glossary of systemic terms they were active participants and researchers. Similarly, the feature 'enhancement' in the clause complex system led to the decision that words like '*embellissement*' or '*accroissement*', which might appear to be good translations for the term 'enhancement', are ill-suited to express logico-semantic relations in French; the final decision went with a term devoid of colour, '*qualification*', but appropriate for the analysis of enhancing relations, such as '*qualification temporelle, causale*' within a text.

The French clause complex system that students work with is formalized in Figure 10.3 and is followed by French definitions of the expansion relations: *élaboration, extension* and *qualification.*

As students tend to talk about French in French in class, they also tend to do their analysis using the French terminology; but some feel more comfortable writing their commentary in English as shown below. The following

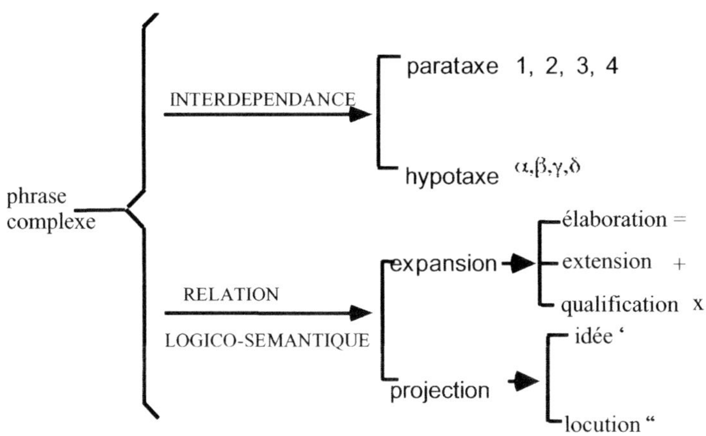

Expansion:
Élaboration: *une phrase en développe une autre en apportant des clarifications, des précisions, en faisant des commentaires ou en donnant des examples.*
Extension*: une phrase en développe une autre en ajoutant de nouveaux éléments, qui peuvent représenter une simple addition (et) ou une exception (ou) ou une différence/contraste (mais).*
Qualification: *une phrase en développe une autre en apportant une qualification temporelle, causale, de lieu, de manière ou de condition.*

Figure 10.3 The French clause complex system

analysed text is an example of the kind of short assignment students perform during the semester which helps build up a resource of analysed texts for the course: they apply one of the analyses they have learnt to a particular text, here a transitivity analysis combining a transitive approach (identifying the different process types) and an ergative approach (identifying functions that generalize across process types: *Agent*, *Véhicule* (Medium), *Etendue* (Range), and then present their results in a summarizing table, followed by a short commentary (in either French or English).

Text 2

Afghanistan : deuxième vague de bombardements (LEMONDE.FR|08.10.01|)

(1) [Agent] Une deuxième vague de bombardements [Matériel: effectif] a frappé Kaboul, [Circ: temporelle] tard dans la soirée de dimanche.
(2) [Véhicule] Les premières operations anglo-américaines [Matériel: moyen] ont commencé [Circ: lieu] en Afghanistan [Circ: temporelle] à 18h15 (heure française).
(3a) [Véhicule] George W. Bush, [Verbal: moyen] a confirmé, [Circ: lieu] dans un discours à la nation [Circ: temporelle] dimanche à 18h50,
(3b) que [Véhicule] les Etats-Unis et la Grande-Bretagne [Matériel: moyen] avaient entamé [Etendue] des frappes militaires 'ciblées'.
(4) [Véhicule] Les forces armées anglo-américaines [matériel: moyen] ont lancé [Etendue] une première offensive [Circ: lieu] sur Kaboul, la capitale afghane, Jalalabad, Kandahar et Hérat.
(5a) [Véhicule] Le porte-parole du ministère pakistanais des affaires étrangères, [Mental: moyen] a regretté [Etendue: metaphenomenon] [[que les efforts diplomatiques n'aient pu permettre de convaincre les dirigeants talibans de 'répondre aux demandes internationales']].
(6) [Agent] Les talibans [relational: attributif] ont qualifié [Véhicule] les bombardements [Etendue] 'd'acte terroriste'.
(7a) [Véhicule] Ben Laden, [Circ: lieu] dans un message préenregistré [[diffusé dimanche soir par Al-Jazira TV]], [relationnel: attributif] se dit [Etendue] prêt à 'la confrontation'
(7b) et [verbal] affirme
(7c) qu' [Véhicule] un 'groupe de musulmans' [matériel: moyen] a bien commis [Etendue] les attentats du 11 septembre.

Resumé des ressources grammaticales (summary of grammatical resources)

A. Procès

	matériel	relationnel	mental	verbal
effectif	1	1		
Moyen	3	1	1	2

B. Participants

Agent	Véhicule	procès	Etendue
Vague de Bombardements		matériel	
	Operations anglo-americaines	matériel	
	Bush	verbal	
	Les Etats Unis et GB	matériel	Frappes militaires
	Les forces armées anglo-americaines	matériel	Première offensive
	Le porte-parole pakistanais	mental	Les efforts diplomatiques
Les talibans	Les bombardements	relationnel	Acte terroriste
	Ben Laden	relationnel	Prêt à la confrontation
	[Ben Laden]	verbal	
	Un groupe de Musulmans	matériel	attentats

C. Circonstances
Lieu: 3
Temporelle: 3

> 'This article, describing the beginning of the bombing in Afghanistan, shows very little usage of Agency. The use of middle constructions decreases the sense of responsibility we feel on the part of the Anglo-American side. On the only occasion where an Agent does carry out a material process, this Agent is the impersonal "vague de bombardements." Such nominalisation of the violent actions described in the article is very prevalent (e.g., Frappes militaires "ciblées", les operations anglo-américaines, les bombardements) and serves to distance the reader from the reality of the events.' (Student's comment)

Students are encouraged from the start to tabulate a quantitative summary of results; it helps them see which grammatical resources have prominence in the text. While Table A on processes and agency reveals that the text has a majority of middle clauses (Agentless), table B further unmistakably foregrounds the lack of Agency. Furthermore, it underlines that the only agent of a material clause is the nominalization, *vague de bombardements* (as expressed in the student's commentary).

This kind of short, highly targeted assignment readily reveals whether the student has understood how grammar is used to construe a particular reality and to convey ideological meaning. In learning to talk about the French language through French, students not only improve their French and their understanding of how French functions; they learn what they can do with the language and the metalanguage and what language can do for them. In other words, they come to realize that learning about language and French in particular is not only a means of learning language and culture but also of making sense of the world and of acting on the world.

The benefits of learning about French in French to advanced learners is well supported by the comments students make in course evaluations (see Appendix A). Not only do they feel their French and understanding of French has improved, they feel that they have acquired skills that are useful to them in other learning contexts, as 'language is implicated in some way or other in all educational activity' (Halliday 1999: 2).

The learning experience of the students in this particular course has shown that learning metalanguage and language in unison is certainly beneficial to advancing second language learning, which would suggest that linguistics should perhaps be an integral part of second language learning. It also indicates that learning language in conjunction with learning something else (rather than in isolation), whether it be linguistics, literature or social sciences, challenges students to use the language in more technical and advanced ways. This is something to consider at a time when language departments are asked to find ways of reducing the amount of courses they offer because the university finds them too costly: it is perhaps time to do away with the old distinction between 'content' and 'language' courses and think of modelling courses along the lines that Byrnes (2001) implemented in the German Department at Georgetown University where all courses are content-oriented language courses:

> The curriculum project, 'Developing Multiple Literacies,' spans all aspects of the Department's 4-year undergraduate curriculum. Taking a content-oriented and task-based approach in all courses, it focuses on content from the beginning of the instructional sequence and gives explicit attention to students acquiring German to levels of performance that are customary in the academy until the time of their graduation. Overcoming both the terminological and substantive dichotomy between language courses and content courses, it arrays instruction in five instructional levels specified by acquisitional goals expressed in terms of content and language learning. (Byrnes, 2001: 518)

This paper focused primarily on advanced learners of French and how learning about French through SFL can enhance their general understanding of language and of French and also their ability to read, write and speak in that language in specialized ways. However, it is important to note that the orientation towards meaning and context of the SFL framework makes it a strong pedagogical tool not only for advancing students' knowledge of how meanings are created in texts in various contexts of use but also for teaching language at any level where the notions of context, register and metafunctions can be used as points of departure for introducing students to use their second language and to mean in different situations. That potential of the SFL framework has led to reconceptualizing as well the instructional approach being used with beginning students in the Department of French Studies at the University of Sydney.

References

Biggs, J. (1999) *Teaching for Quality Learning at University*. Buckingham: The Society for Research into Higher Education and Open University Press.

Byrnes, H. (2001) 'Reconsidering graduate students' education as teachers: "It takes a department!" '. *The Modern Language Journal*, 85, 512–30.

Caffarel, A. (2006) *A Systemic Functional Grammar of French: From Grammar to Discourse*. London: Continuum.

Christie, F. (1983) 'Learning to write: a process of learning how to mean'. *English in Australia: Journal of the Australian Association for the Teaching of English*, 66, 4–17.

Christie, F. (ed.) (1990) *Literacy of a Changing World*. Melbourne: Australian Council for Educational Research.

Christie, F. and Rothery, J. (1979) *Language in Teacher Education: Child Language Development and English Language Studies*. Applied Linguistics Association of Australia (Occasional Papers).

Christie, F. and Unsworth, L. (2005) 'Developing dimensions of an educational linguistics', in R. Hasan, C. M. I. M. Matthiessen and J. Webster (eds), pp. 217–50.

Decaux, A. (1987) *Alain Decaux raconte l'Histoire de France aux enfants*. Paris: Librairie Académique Perrin.

Halliday, M. A. K. (1978) *Language as Social Semiotics. The Social Interpretation of Language and Meaning*. London: Edward Arnold.

Halliday, M. A. K. (1979) 'Differences between spoken and written language: some implications for literacy', in G. Page *et al.* (eds), *Communication through Reading*. Adelaide: Australian Reading Association.

Halliday, M. A. K. (1985) *Spoken and Written Language*. Geelong, Australia: Deakin University Press.

Halliday, M. A. K. (1994) *An Introduction to Functional Grammar*. 2nd Edn. London: Edward Arnold.

Halliday, M. A. K. (1999) 'The notion of "context" in language education', in M. Ghadessy (ed.), *Text and Context in Functional Linguistics*. Amsterdam/Philadelphia: John Benjamins, pp. 1–24.

Halliday, M. A. K. (2002) 'Grammar and daily life: concurrence and complementarity', in Jonathan Webster (ed.), *On Grammar*. London: Continuum, pp. 369–83.

Hasan, R. and Martin, J. R. (eds) (1989) *Language Development: Learning Language, Learning Culture*. (Volume 1 of *Meaning and Choice in Language: Studies for Michael Halliday*.) Norwood, NJ: Ablex.

Hasan, R., Matthiessen, C. M. I. M. and Webster, J. (eds) (2005) *Continuing Discourse on Language: A Functional Perspective*, Volume 1. London: Equinox.

Kress, G. (1982) *Learning to Write*. London: Routledge & Kegan Paul.

Lyons, J. (1968) *Introduction to Theoretical Linguistics*. London: Cambridge University Press.

Martin, J. R. (1985) *Factual Writing: Exploring and Challenging Social Reality*. Geelong, Australia: Deakin University Press (Sociocultural Aspects of Language and Education).

Martin, J. R. (1986) 'Intervening in the process of writing development', in C. Painter and J. R. Martin (eds), *Writing to Mean: Teaching Genres across the Curriculum*. Applied Linguistics Association of Australia (Occasional Papers 9), pp. 11–43.

Martin, J. R. and Painter, C. (eds) (1986) *Writing to Mean: Teaching Genres across the Curriculum*. Applied Linguistics Association of Australia (Occasional Papers 9).

Martin, J. R. and Rose, D. (2005) 'Designing literacy pedagogy: scaffolding democracy in the classroom', in R. Hasan, C. M. I. M. Matthiessen and J. Webster (eds), pp. 251–80.

Melrose, R. (1995) *The Communicative Syllabus, A Systemic-Functional Approach to Language Teaching*. London: Pinter.

Rowland, S. (2000) *The Enquiring University Teacher*. Buckingham: The Society for Research into Higher Education and Open University Press.

Appendix A: Selected student comments about their learning experiences in the course

'very interesting and helps French as well as English grammar.'

'Linguistics now seems much more relevant to me and is certainly useful for communication skills in a BA.'

'A broader perspective on linguistics and reality etc helped me in French and sociology (cultural studies & semiotics etc.).'

'grammatical skills, analysis, representation of reality, greater understanding of language and components.'

'Definitely a greater appreciation and understanding of why writers make choices in language.'

'very interesting and helpful for other French (and English!) courses that I am taking.'

'it has helped me to analyse texts I am using in other subjects. It has also helped me write essays and articles in English (edit them to increase the lexical density and therefore fit more meanings into the clause).'

'I found the subject very interesting as it could be applied and applied to both English and French as we were given examples in both. A good overall introduction to linguistics for non-linguists.'

'better appreciation of language and its uses in respect to meaning and structure.'

'The course is very lively-useful, good either for English or French. It does establish good understanding of the French language with enough depth on the functions of the language.'

'I have learnt the basics for French linguistics which will help me in doing Textual Linguistics next semester as I have enjoyed this subject so much and wish to study linguistics further.'

'Good to have all classes in French. Helps us improve our French.'

'I found the course interesting. I have never done linguistics before so I feel as though I have learnt a lot. It has given me a new perspective on the function of language & how we can be manipulated by it. I have developed critical thinking skills.'

'it has helped my French reading skills.'

'it gave me a new way of looking at texts.'

'It was an interesting way of looking @ french & understanding it better.'

'The analytic skills I have developed in this unit have already helped me in other subjects and I use them everytime I read a newspaper! Very useful for any language study and especially for writing essays. My own writing is more concise and directed now.'

'To look at texts in a different way – eg to see how seemingly neutral words and phrases can carry hidden opinions.'

'Improve, consolidate my comprehension skills in French.'

'A much better understanding of the use of the French language.'

'I have learnt to critically analyse all pieces of writing rather than taking it at face value.'

'This course gave me a new way of looking at texts both in English and French, in interpreting the devices used by the author to influence the reader's opinion.'

'Different ways of reading and thinking about a text – I've noticed how, with other subjects (not just French) how I look at a text or attempt to analyse it has been influenced by what we've been doing throughout this semester.'

'better awareness of text, context, use of textual effects dependent on situation and aim of writer.'

'I learnt lots of analytical skills that can be applied in different areas.'

'Wide range of skills are assessed.'

'I think to learn about grammar is kinda interesting, as it casts light on how we construct meaning, both in our native language and in foreign languages. It's interesting to observe the similarites and differences between how grammar is used and how meaning is constructed in different languages and cultures. Studying grammar teaches us a lot about our own language, even when studying the grammar of another language. We can learn to appreciate the complexities of our language, and gain a better understanding of how the words we use, and how we use them can be interpreted.'

'It is good to have a linguistics "department" within the French department. Keep up the good work, Alice.'

Part III

Programmatic and Curricular Issues

11 Modelling a genre-based foreign language curriculum: staging advanced L2 learning

Cori Crane

In recent years, foreign language (FL) educators have begun to place increasing importance on meeting the needs of advanced-level second language (AL2) learners in instructional environments. This shift in awareness has led to the discipline's acknowledgement that support of upper levels of language development cannot merely be addressed through individual efforts, whether these come from teachers or from students, but must be carefully planned over longer instructional sequences in order to assure successful outcomes at various stages along a developmental continuum that is inherently long term. For that reason, a curricular environment provides a relevant site for discussing the notion of advancedness, particularly when it also considers multiple endpoints and appropriate means for reaching them, e.g., through materials, pedagogies and assessment. Discussions of L2 learning have also occasioned the reconceptualization of key notions regarding language and language development. Scholars who have considered the linguistic needs of AL2 learners within instructed environments (e.g., Byrnes and Maxim 2004; Byrnes *et al.* 2006; Swaffar and Arens 2005) have tended to frame their arguments within the constructs of literacy, discourse and genre for these constructs' ability to address the social constructedness of language at higher levels.

Although the linguistic profile of the AL2 learner is still emerging and much research is needed to enrich our understanding of the needs and abilities of this heterogeneous learner group, diverse social contexts across a range of public, professional and institutional settings have been identified as pointing to what we generally call advanced language use. Expressed in terms of genres (e.g., abstracts, reports, reviews, summaries), these forms of language use with their preference for abstract thought and multiple other-oriented viewpoints have increasingly been characterized in terms of linguistic features whose occurrence is highly probable in those settings (see Byrnes and Maxim 2004; Byrnes *et al.* 2006; Johns 2002; Schleppegrell and Colombi 2002). Particularly prominent are the resources of grammatical metaphor (Christie 2002; Martin 2002b), complex clause-combining strategies (Colombi 2002), and an expanded repertoire of evaluative resources

(Martin 2002b; Schleppegrell 2002) (for how these features characterize AL2 writing, see Ryshina-Pankova, this volume).

Building on this research, Byrnes and Sprang (2004) conceptualize an L2 curricular continuum where a learning pathway from the *primary discourses* of familiar life to the *secondary discourses* of public institutions (Gee 1998) correlates with dominant linguistic realizations and underlying contextual factors that fall along the three variables of *field, tenor* and *textuality* (cf. *mode*) that Systemic Functional Linguistics (SFL) uses to describe different registers and their variations (see Matthiessen 1993). Their continuum is not meant to follow a strict linear developmental path; rather, it represents a range of discourses that are used recursively, building and expanding on previous linguistic patterns and their contexts of use.

Consideration of how register progresses across a curriculum can provide important insights into the relationship between linguistic features and their situational contexts. How these variables unfold in a text, however, is informed by a second semiotic system, that of genre. For curriculum design, the twin constructs of context in the SFL tradition, i.e., register as 'context of situation' and genre as 'context of culture' (Halliday 1999), provide a useful lens for seeing how configurations of linguistic phenomena construe meanings across texts. These two constructs do not function independently of each other, but refer to different proximities between the observer and the linguistic phenomena being observed. That is, the register angle provides information about form–meaning relationships at the situational level in terms of the ideational (content), interpersonal (participants and their relationship to each other), and textual metafunctions (contribution that text plays in presenting information) (Halliday 1985), whereas the genre perspective considers how the larger meaning potential that these register variables take up map onto preferred global text structural patterns that cultures typically use to accomplish particular communicative purposes (Hasan 1984/1996; Martin 1992). As Martin (1985: 250) explains, 'Genres are how things get done, when language is used to accomplish them.'

As 'a staged, goal-oriented social process' (Martin 1984), genre embodies three fundamental aspects of language use that make the construct especially applicable for envisioning L2 curricula: communicative purpose, stagedness and social embeddedness. Communicative purpose is a fundamental property of all genres, and linguistic choices made within a given genre presumably work towards accomplishing its particular communicative goal. In order to successfully realize a given communicative action, genres draw on various obligatory and optional 'verbal strategies' (Martin 1985) that represent their own micro-level communicative purposes while contributing to the text's overall message. These stages, or 'moves' (cf. Swales 1990), reflect the global organizational patterns that texts reveal as they unfold, and it is in their unfolding, not in the properties of the text alone, that communicative purpose is achieved. This configuration of obligatory and optional stages is referred to as the genre's schematic structure. Hasan's (1984/1996) construct of 'Generic Structure Potential' presents a useful model for understanding the total range of possible 'structural resources' that a given genre can draw on for its communicative purpose to be achieved successfully within a given social context. Stages can appear recursively throughout a text and need not follow in linear fashion (cf. the non-linear, nuclear schematic structure of the hard news story, White 1997). With its emphasis on options, the model provides ample flexibility for expression of thought and acknowledges that genres, while relatively stable cultural entities precisely because they are born out of rather stable, recurring situations in which people interact (Bakhtin 1986; Kress 1993), can transform into new generic forms recognizable within a given culture.

Knowledge of genre and generic structure has proven to be a useful pedagogical resource for developing L1 and L2 learners' reading and writing abilities (see Johns 2002). At the curricular level, the construct of genre has been explored to understand the dominant ways of meaning-making that educational programs privilege (Macken *et al.* 1989; Swales 1990), to see the preferred ways in which learners themselves choose to represent themselves (Rothery and Stenglin 1997, 2000), and to chart the developmental learning path that students follow in becoming socialized within particular disciplines, such as in secondary-school history (Coffin 1997; Martin 2002a, 2003). Mapping out the genres that learners progress through across a curricular environment and considering how these genres relate to each other structurally and linguistically allows educators to see the relationship between language and content, connections that are otherwise difficult to detect when linguistic units, functional categories or communicative situations alone act as the guiding constructs for curriculum design.

In this chapter, I explore how knowledge of genre, specifically that of generic structure, can aid L2 curriculum designers in sequencing and supporting instructional materials to facilitate L2 learners' comprehension and production along various points in their language development (see also Byrnes 2006 and this volume). In line with the focus of this volume, my interest is in exploring how curriculum and instruction might support the attainment of high levels of L2 abilities. However, as already stated, that kind of support is likely to require considerable staging much earlier than the courses that are typically designated as 'advanced'. For this reason, I focus on how two story genres that can occur at various curricular levels, the *recount* and the *personal narrative*, will be pedagogically enacted quite differently at different curricular levels and yet will also show important links. By foregrounding the relationship among different genres and by showing their similarities and differences I hope to provide a crucial perspective on curricular design that uses genre as a central construct. I argue that a full understanding of the generic properties of texts and their pedagogical value must encompass how they might scaffold each other at different instructional levels, particularly if learners that traverse beginning and intermediate instructional levels are ultimately to become successful language users at the advanced levels.

Two texts representing beginning and early-advanced levels within the undergraduate curriculum of the Georgetown University German Department (GUGD) serve as sites for investigating how different linguistic resources map onto different schematic structures within a projected progression of genres across a curriculum. As engagement with these texts fosters intricate links between textual meaning and language forms, it also provides rich opportunities for engaging with distinct aspects of the L2's cultural practices and values, an ability that surely ranks as one of the most desired characteristics of advanced levels of ability (see a similar discussion in Ryshina-Pankova, this volume).

Multiple literacies and narrativity

Over a three-year period, from 1997 to 2000, the GUGD revamped its undergraduate curriculum to create stronger articulation between lower and higher division courses and to promote learning at all stages of L2 development through content-based instruction, a model that deliberately sought to break away from the traditional, bifurcated system of separate language and content courses typically found across US college-level FL departments (see Byrnes 2001 for further discussion on the curriculum renewal project).

The programmatic changes surrounding this revision process, which included the decision to do away with commercial textbooks beyond the beginning level and to create instructional materials that reflected a commitment towards developing L2 literacy abilities across various personal and public social contexts, led to the constructs of text and genre taking on increasing importance across all aspects of curriculum design, i.e., articulating goals and objectives, developing and sequencing instructional materials, creating and supporting pedagogies, and informing assessment. Selection and didacticization of materials, as well as the creation and instructional support of writing and speaking tasks, considered central components of GUGD's curriculum, relied on a detailed understanding of the semiotic resources typically found across text types moving from primary to secondary discourses.

Although the new curriculum was designed to encompass a broad palette of genres and discourses, as its name 'Developing Multiple Literacies' implies, from its inception, narrativity was given a prominent place in conceiving L2 development from the beginning to advanced instructional levels.[1] Because narratives permeate both the primary and secondary discourses in FL college-level programs, they are ideal text structures to consider in curricular sequencing. Two key features of stories in particular stand out as offering a wide range of lexicogrammatical choices that can be realized in more and less complex ways: (1) spatial and temporal displacement across a series of events realized through such features as tense and temporal and spatial adverbial expressions, and (2) character development and point(s) of view realized through such strategies as reported speech and modality (Ochs 1997). Furthermore, narratives have long been noted as presenting rich information about cultural values, behaviours and practices. As vehicles for representing stories of the past or imagining worlds of things that might be, narratives provide important resources for making sense of the world and can therefore serve as especially useful tools in the development of content knowledge.

Across the GUGD curriculum, the preference for narrativity can be seen particularly clearly in the writing tasks located across the first four instructional levels (see Table 11.1). All tasks in bold represent genres in which telling a story or recounting a series of events is a central communicative goal towards completion of the task.

From this table, it becomes clear that few tasks involve students writing pure stories. Instead, narrative structures tend to be embedded within larger generic frameworks – epistolary genres at the lower levels and journalistic discourse at intermediate and advanced levels. This makes sense given that personal letters, genres that L2 learners are likely to have been exposed to in their L1s, tend to mark specific participants and communicative purposes explicitly in texts, making them well-motivated contexts within which L2 writers might frame their stories. In the newspaper reports, narration typically manifests itself via 'human interest stories', where personal stories are recounted to show the significance of particular socio-political phenomena.

Modelling the *Recount*: beginning-level L2 instruction and the construal of cultural practices

In the GUGD's beginning-level German course 'Contemporary Germany', stories are primarily used to convey information about German cultural practices. The genre *recount* provides an appropriate textual environment

Table 11.1 Genres represented among writing tasks across the GUGD undergraduate curriculum

Level I (Beginning)	Level II (Intermediate)	Level III (Advanced)	Level IV ('Text in Context')
1. **Personal letter**	1. **Newspaper report**	1. **Thank-you letter**	1. **Condolence letter**
2. Invitation	2. **Newspaper report**	2. **Personal narrative**	2. **Newspaper report**
3. **Personal letter**	3. Manifesto	3. **Narrative summary**	3. Letter-to-the-Editor
4. **Personal letter**	4. Resumé	4. Manifesto	4. Précis
5. **Postcard**	5. **Fairy tale**	5. Newspaper report	5. Précis
6. **Official letter (police report)**	6. **Novel excerpt**	6. **Newspaper report**	6. Précis
7. **Personal narrative**			7. Formal speech
8. Horoscope			

through which such information can be interpreted on a personal level. In this section, I investigate the typical textual and linguistic features of this story genre in order to show the opportunities for developing L2 writing abilities at this level and to point to the sorts of cultural meanings that the text type can construe.

The recount's episodic, linear schematic structure reflects the text's communicative purpose of conveying a 'journey' across a series of temporally organized events. Because there is no high point, or conflict, there is also no resolution. The point of view remains constant, as participants tend not to shift roles in the transitivity patterns, which consist mainly of material process verbs connoting action (Rothery and Stenglin 1997). The recount comprises two obligatory stages: (1) an Orientation, where setting and participants are introduced, and (2) a Record of Events stage, in which experiences are relayed in consecutive temporal sequence. An optional Reorientation stage concludes the genre, where events return to the starting point of the story, often with an interpersonal slant. Additional optional stages of Abstract, Synopsis and Coda create explicit bridges to the story and conversation worlds and are more commonly found across oral stories. Figure 11.1 shows the recount's schematic structure and the communicative questions that its stages answer:

(Abstract):	What is this story about? (interpersonally based evaluation)
(Synopsis):	What is this story about? (experientially based summary)
Orientation:	Who, when, what, where?
Record of Events:	What happened?
(Reorientation):	How did it end and how does it relate to the beginning?
(Coda):	What does the story look like now? (bridges story and conversation times)

Figure 11.1 Schematic structure of recount
(based on Rothery and Stenglin 1997; Macken-Horarik 2002)

Text 1 *Am Brandenburger Tor in Berlin* ('At the Brandenburg Gate in Berlin') (Hieber 1998) is a recount used in the fourth thematic unit of 'Contemporary Germany' on holidays and celebrations, where past tense verb forms as well as time expressions used to describe daily routines are targeted. The text relays events of a German teenager's New Year's festivities in the city of Berlin and answers the experientially based question: How did you celebrate New Year's Eve? The recount's journey metaphor makes it ideal for construing experience of cultural traditions. The text refers not only to activities surrounding New Year's Eve, i.e., watching fireworks, club-hopping, parties and eating customs among young people, but also provides information about the city, including forms of transportation.

Text 1: Recount: *Am Brandenburger Tor in Berlin* ('At the Brandenburg Gate in Berlin')

Orientation	*In Berlin wohnen Binny und Steffi, zwei Freundinnen von mir. Die kenne ich noch aus der Schule. Zusammen mit meiner Freundin Conny und ihrer Mutter sind wir dorthin gefahren, von Dresden aus. Binny und Steffi haben uns in Berlin abgeholt. Einen Tag später sind noch weitere Bekannte nachgekommen, die in Leipzig übernachtet haben. Die sind von unserer Tanzgruppe. Wir haben sie auch vom Bahnhof abgeholt.*	Binny and Steffi, two girlfriends of mine live in Berlin. I know them from school. Together with my friend Conny and her mother, we drove there from Dresden. Binny and Steffi picked us up in Berlin. A day later, still more friends followed, who stayed over in Leipzig. They're from our dance troupe. We picked them up from the train station as well.
Record of Events	*Zuerst waren wir in der Wohnung und habe da gefetet und Party gemacht und noch Raclette gegessen. So bis um halb zwölf. Dann sind wir los in die Wohnung von Bekannten von Binny und Steffi und von dort wie Katzen über eine Leiter hoch aufs Hausdach geklettert. Da standen wir – Luftlinie so 300 Meter vom Brandenburger Tor entfernt – hoch über Berlin, zwischen den Schornsteinen.*	First, we were in the apartment and partied there [and partied] and ate raclette, too. Till half past 11. Then we took off to the apartment of friends of Binny and Steffi and from there climbed over a ladder like cats high onto the rooftop. There, we stood – a beeline of 300 meters away from the Brandenburg Gate – high above Berlin, among the chimneys.
	So gegen 1 Uhr sind wir wieder vom Dach runter, sind dann in einen Club. ***Das war nicht so toll, weil die Musik ziemlich schlecht war und die Leute so komisch.*** *Dann sind wir so gegen drei oder halb vier wieder zu Hause gewesen.*	Around 1 o'clock, we came down again from the roof, then went into a club. **That wasn't so great, since the music was pretty bad and the people so strange.** Then around 3 or 4, we were back at home.
	Am Freitag, dem Neujahrstag, haben wir uns ausgepennt, sind erst so um 4 Uhr aus dem Haus, zum Potsdamer Platz, haben uns da die Arkaden angeguckt. Abends noch die Oranienburger Straße in ein mexikanisches Restaurant. Und danach wieder, zusammen mit anderen Bekannten von Binny und Steffi, in einen Club. Die ganze Nacht voll durchgemacht. Dann nach Hause um 4 Uhr. Bis 7 Uhr morgens saßen wir mit den Jungs noch in der Küche und haben Kaffee getrunken. Bis sie uns um halb 9 so zum Bahnhof gebracht haben.	On Friday, New Year's Day, we slept in, got out of the house finally around 4 o'clock, direction Potsdam Square, took a look at the arcades there. In the evening in a Mexican restaurant on Oranienburger Street. And then afterwards, together with other friends of Binny and Steffi to a club. Stayed up the whole night. Then back home at 4 o'clock. Till 7 o'clock in the morning, we sat with the boys in the kitchen and drank coffee. Till they brought us to the train station at half past eight.

Reorientation	***Ein tolles Silvester! Das beste, das ich je hatte! Wirklich supergut!***	A great New Year's Eve! The best one I've ever had! Really awesome!

With few exceptions, the entire text is comprised of simple clauses that convey information on processes, participants involved in the processes and circumstantial adjuncts of temporal and spatial location. Textual coherence is created primarily through personal reference chains, as well as temporal conjunctions. The unfolding of events construed through these simple transitivity patterns contributes to the sense of harmony and commonality of experience typical of recounts (Rothery and Stenglin 1997), leading to a sense that the New Year's festivities are overall enjoyable.

A look at the generic stages of the recount points to slight differences in how the story unfolds linguistically. The opening Orientation stage draws on relational, mental and material process verbs (bolded) to present the participants and setting:

> *In Berlin* ***wohnen*** *[relational] Binny und Steffi, zwei Freundinnen von mir. Die* ***kenne*** *[mental] ich noch aus der Schule. Zusammen mit meiner Freundin Conny und ihrer Mutter* ***sind*** *wir dorthin* ***gefahren*** *[material], von Dresden aus.*
> Binny and Steffi, two girlfriends of mine ***live*** [*relational*] in Berlin. I ***know*** [*mental*] them from school. Together with my friend Conny and her mother, we ***drove*** [*material*] there, from Dresden.

In the Orientation, the participants are also primarily located in the thematic position, anchoring their centrality in the text. Actions are construed through the past tense; elaborations of the participants and setting are expressed through the present.

In the Record of Events stage, two grammatical patterns emerge: (1) material process verbs (e.g., 'partied, ate') in the past tense that connote activities the individuals engage in, and (2) thematic positioning of temporal adverbs that situate these actions against a timeline. Also found across this stage are ellipsed clauses found towards the end of the recount, where actors (e.g., 'we') and auxiliary verbs (e.g., 'had/were') are omitted:

> *Abends noch die Oranienburger Straße in ein mexikanisches Restaurant. Und danach wieder, zusammen mit anderen Bekannten von Binny und Steffi, in einen Club. Die ganze Nacht voll durchgemacht.*
> In the evening in a Mexican restaurant on Oranienburger Street. And then afterwards, together with other friends of Binny and Steffi to a club. Stayed up the whole night.

Absence of this given information underscores the strong sense of sequential progression characteristic for recounts. Evaluative comments (see Text 1, in bold) about the night appear across the Recorded Events and in the Reorientation stages, construed through relational process verbs (e.g., 'was') and appraisal attributes of appreciation (e.g., 'great', 'awesome').

In teaching this text to beginning-level L2 German students, attention is placed on connecting the major lexicogrammatical features, i.e., material process verbs that refer to travelling and celebrating, and temporal expressions in thematic position, to the story's unfolding meaning in terms of the three stages. In class discussions, instructors encourage students to view the text as emblematic of larger cultural traditions of German youth and ask students to draw cultural comparisons to their own New Year's celebrations.

This discussion culminates in a writing task where students are given a more formal occasion to engage with the genre with regard to its typical linguistic forms and overall message. In a personal letter, students are asked to explain to a German penfriend how New Year's Eve is celebrated in the United States and to provide an example through a short recount of their own holiday experiences. In the body of the letter, where discussion of holiday customs and the embedded recount is to be located, students are encouraged to draw on similar features used in the source text, including: (1) temporal expressions in the thematic position of clauses, (2) past tense of primarily material process verbs, and (3) attributes functioning as appraisal resources to evaluate the event(s). As with all the writing tasks across the curriculum, this connection between stages and expected linguistic forms is outlined explicitly in task instructions provided to the students and supported in instruction leading up to the writing event.

Modelling the Personal Narrative: advanced-level L2 instruction and the construal of cultural values

Turning now to advanced-level instruction, narrative continues to play a prominent role and is, in fact, the thematic focus of the GUGD's sequenced fifth- and sixth-semester course, 'German Stories, German Histories'. The course, representing the first of three advanced instructional levels within the curriculum,[2] uses personal and public narratives as a means for exploring modern German history (see also Eigler 2001 for further description of this course). Both instructional materials and tasks for the course draw on various genres that contain narration as a guiding rhetorical structure, e.g., short story, autobiographical account and historical recount. Personal narrative, however, dominates among the genres in the course.

In contrast to the personal recount, where episodic unfolding of events correlates to a sense of common, shared experience for the purpose of informing or entertaining, the personal narrative involves a crisis moment set against a series of events that attempts to make sense of the problematic experience. Both the conflict and the resolution provide a window into the values of a given culture – particularly in terms of what is perceived as a conflict or disruption, as well as how a member of a given culture responds to that challenge. Through the experience, participants, whose roles typically shift between acting and being acted upon, interact with the experience through multiple vantage points, often revealing a complex set of perspectives.

The text at the centre of this discussion is the personal narrative, *Drei Freunde* ('Three Friends') (Klecker 1991). Set against the mounting political tensions surrounding the Berlin Wall, the narrator tells the story of two friends who, over the years growing up in East Germany, develop different viewpoints on their civic role in the country. One friend, Wolfgang, escapes to the West through extracting information from the other friend Eberhardt, a loyal border patrol soldier. Shots are fired during the escape, and though Wolfgang makes it to the other side, his accomplice does not and is killed by Eberhardt. The narrator, sympathetic to Wolfgang's situation, decides he can no longer continue his friendship with Eberhardt.

A moves analysis of 'Three Friends' reveals that its schematic structure corresponds to the stages identified in analyses of (oral) personal narrative (Labov and Waletzky 1967; Martin and Plum 1997; Rothery and Stenglin 1997). Figure 11.2 shows the optional and obligatory stages of the genre and the communicative questions each stage answers.

(Abstract):	What is this story about? (interpersonally based evaluation)
(Synopsis):	What is this story about? (experientially based summary)
Orientation:	Who, when, what, where?
Complication:	Then what happened?
Evaluation:	So what?
Resolution:	What finally happened?
(Coda):	What does the story look like now? (bridges story and conversation times)

Figure 11.2 Schematic structure of personal narrative

Similar to the New Year's recount, the personal narrative 'Three Friends' unfolds temporally. In contrast to the recount, however, the narrative events are not merely connected via coordination of successive events but are referenced through subordinating conjunctions, nominalized events, prepositional phrases and temporal adverbs that serve to locate the setting, portray reoccurring events, and show how events relate to each other both temporally and causally. These temporal resources also tend to cluster temporality at the beginning of many of the stages, making the schematic structure immediately transparent. Figure 11.3 displays the schematic structure of the text along with the explicit temporal expressions found across each stage. In the following section, I present a brief description of the most salient linguistic features that characterize the opening and concluding stages of this narrative.

The text-incipient stages: Abstract and Orientation

The Orientation stage consists of three distinct thematic foci that lay out the personal and political contexts necessary for understanding the story: descriptions of (1) the men's friendship as boys, (2) the political context in East Germany and (3) the men's friendship as adults. The first Orientation section, in which the friendship of the three boys is described, resembles a personal recount: relational process verbs (e.g., 'were', 'is') first introduce the characters in an Orientation stage, a Record of Events realized through material process verbs (e.g., 'played', 'went', 'fell in love') then follows and a Reorientation that recontextualizes the relationship between the three men in terms of similarities and differences concludes the scene (see Text 2).

Orientation #1 [Boyhood friendship]	**Conjunction**: als ('when')
Abstract	**Adverb of temporal location**: nun ('now')
Orientation #2 [Historical setting**]**	**Nominalization of event**: durch den Bau der Mauer ('through the building of the Wall')
Evaluation	**Nominalization of event**: Der Bau der Mauer ('the building of the Wall')
Orientation #2 **[**Historical setting**]**	**Conjunction**: nachdem ('after')
Orientation #3 **[**Adult friendship**]**	**Preposition of temporal location**: im Jahre 1968 ('in 1968') **Conjunction to signal usuality**: wenn ('when') **Adverbs of duration**: immer ('always'), nach und nach ('little by little')
Complication	**Adverbs of temporal location:** dann, eines Tages ('then, one day'), später ('later') **Conjunction**: als ('when')
Resolution	**Adverbs of temporal location**: auf einmal ('all of a sudden')
Evaluation	---
Resolution	**Adverbs of temporal location**: in der folgenden Zeit ('in the time following, thereafter'), schließlich ('finally') **Conjunction**: als ('when') **Adverbs of duration**: nicht mehr ('no longer'), immer mehr ('more and more'), immer seltener ('less and less')
Coda	**Adverb of temporal location**: heute ('today') **Preposition of temporal location with nominalization of event:** nach der Wende ('after the fall of the Berlin Wall')

Figure 11.3 Schematic structure of *Drei Freunde* ('Three Friends') with linguistic features of temporality

Text 2: *Drei Freunde* ('Three Friends'): Opening Orientation and Abstract

Orientation in Personal Narrative	**Recount Orientation:** *Wir* ***waren*** *drei Freunde. Keine besonders dicken Freunde, aber eben doch gute Freunde, wie das unter Schuljungen üblich* ***ist****.*	**Recount Orientation:** We **were** three friends. Not especially close friends, but still good friends, as **is** common among schoolboys.
	Recount Record of Events: *Wir* ***spielten gemeinsam*** *Fußball,* ***gingen gemeinsam*** *ins Kino und* ***trieben gemeinsam*** *Unfug. Später, als wir älter* ***wurden, fuhren*** *wir im Urlaub* ***gemeinsam*** *zelten,* ***gingen gemeinsam*** *zum Tanz und* ***verliebten*** *uns in die* ***gleichen*** *Mädchen.*	**Recount Record of Events:** We **played** football **together, went** to the movies **together** and **got into trouble together**. Later, as we **became** older, we **took** camping vacations **together, went** to dances **together** and **fell in love** with the **same** girls.

Abstract in Personal Narrative	**Recount Reorientation:** *Nun sollte man meinen, dass es bei so viel* ***Gemeinsamkeit*** *kaum zu nennenswerten* ***Meinungsverschiedenheiten*** *kommen kann. Aber gerade in einem Punkt gab es* ***unterschiedliche Auffassungen****, und zwar zur politischen Situation.*	**Recount Reorientation:** Now, one should think that with so **much in common** there would hardly be any **differences in opinion** worth mentioning. However, exactly in one area there were **different views**, namely concerning the political situation.

The similarity between the friends topicalized in this opening passage provides a contrastive reference point for anticipating in the next section how political decisions like the building of the Berlin Wall could tear families and friendships apart. The transitivity patterns of material process verbs in simple clauses and the repetition of the attribute *gemeinsam* ('together') contribute to the sense of a balanced world that makes identification with the text by a broad audience possible.

This idyllic picture abruptly ends with the next narrative stage, the Abstract. Here, the previous boyhood memories are referenced through the nominalization *Gemeinsamkeit* (translated here as 'much in common'), whose attribute root *gemeinsam* (translated here as 'together') appears repeatedly throughout the previous Orientation section. In the Abstract, the unfolding commonality of experience between the boys is interrupted by the presence of two nominalizations that are meant to reframe the three men's relationship to each other, one that is now marked by *Meinungsverschiedenheiten* ('differences in opinion') and *unterschiedliche Auffassungen* ('different views'). In this way, the Abstract, whose communicative function is to introduce the story with evaluative commentary, prepares the reader for conflict and struggle to come.

Immediately following the Abstract, the political situation is presented in a second Orientation section that deals with historical events surrounding the story (see Text 3).

Text 3: *Drei Freunde* ('Three Friends'): Orientation #2 [Historical Setting]

Orientation [Historical setting]	***Durch den Bau der Mauer*** *dokumentierte die kommunistische Regierung der DDR für alle Welt sichtbar, dass sie die Spaltung Deutschlands für alle Zeiten als gewollt betrachtet.*	**Through the building of the Wall**, the communist government of the GDR [East Germany] visibly documented for the entire world that it viewed **Germany's division** as intended for all times.
Evaluation	***Der Bau der Mauer*** *war ein solches Beispiel dafür, dass die Mächtigen, wenn sie nicht über ausreichende Argumente der Vernunft verfügen, hemmungslos ihre Macht einsetzen, um* ***politisch Andersdenkende*** *zu disziplinieren.*	**The building of the Wall** was such an example for how the powerful when they do not possess sufficient reasonable arguments wield their power unscrupulously in order to discipline **political dissenters**.

Orientation [Historical setting]	*Nachdem die Mauer gebaut war, konnte in der DDR die Wehrpflicht als Gesetz eingeführt werden. Junge Männer, die keine Lust hatten, diesem verlogenen Staat zu dienen, hatten nun keine Möglichkeit mehr, in die BRD auszuweichen. Aber es gab auch junge Männer, die glaubten, dass das Ehrenkleid, wie die Uniform blasphemisch genannt wurde, anzulegen, ihre patriotische Pflicht sei.*	After the Wall was built it was possible in the GDR to establish compulsory military service as law. Young men who had no desire to serve this hypocritical state now no longer had the opportunity to slip away to the FRG [West Germany]. But there were also young men who believed that putting on the dress of honour, as the uniform was blasphemously called, was their patriotic duty.

This section, in contrast to the verbal, congruent Orientation stage before it, is marked by nominal groups that reference historical events (e.g., 'the building of the Wall, Germany's division') and participants (e.g., 'political dissenters') tied to the post-war Germanies within a setting of subordinate clause complexes. The effect is a tight, dense depiction of these events with room for the narrator's critical evaluation of the East German state. With this shift in content focus and linguistic form, this Orientation section resembles key linguistic features found in the genre of historical recount, where a series of events are 'packaged' into chunks of historical periods and phases (Martin 2002b).

The Resolution: tracking voices and making sense of the events

As is typical for personal narratives, voice plays an integral role in this story. In addition to the narrator's own distinct evaluations of the events, the story is pieced together through information provided to him via media reports, friends and acquaintances, and letters. The Resolution stage, in which these voices emerge most prominently, serves to make sense of the complicating actions leading up to the escape attempt. Due to space constraints, only excerpted examples from this stage are included.

Various linguistic resources track the different sources of information and points of view across the Resolution stage: (1) indirect speech patterns, (2) reported speech marked through reporting verbs and projected clauses and (3) nouns with post-modifiers. In German, indirect discourse can be encoded in the verb to denote that reported speech does not belong to the speaker or represents situations where the speaker wishes to question or distance himself from a particular message. In Example (1), the grammaticized indirect speech marking '*sei*' ('is') allows the narrator to mock Eberhardt's words and to set up a more explicit evaluation regarding Eberhardt's boasting behaviour in the sentence that follows:

(1) Eberhardt erzählte nur, dass er für heldenhaftes Verhalten im Dienst ausgezeichnet worden ***sei***. [cf. direct speech form: *ist*] *Diese Prahlerei konnte er sich einfach nicht verkneifen.*

Eberhardt related only that he **had been** distinguished for his heroic conduct in the line of duty. From this boasting, he simply could not refrain.

Reported speech is also marked repeatedly in the story through reporting verbs and projected clauses to capture conversations and confrontations:

(2) ***Als er mich dann direkt fragte, warum ich seine Gesellschaft meide,*** *konnte ich nur antworten, dem Umgang mit Helden nicht gewöhnt zu sein.*
When he then asked me directly why I was avoiding his company, I could only answer that I was not used to dealing with heroes.

Finally, nouns with post-modifiers that reference reported speech from earlier accounts in the story provide an additional means for indicating authorial voices in the text:

(3) Als sie jedoch ***der Aufforderung Eberhardts, sich festnehmen zu lassen,*** *nicht nachkamen, schoss Eberhardt auf seinen alten Freund.*
However, when they did not meet **Eberhardt's demand to let themselves be apprehended**, Eberhardt shot his old friend.

This nominal structure is further complicated by the embedded infinitival construction to mark the intentions behind the 'demands'. Through such complex lexicogrammatical structures as indirect speech, complex clause-combining strategies and grammatical metaphor, this text is able to present and reference multiple voices into a coherent whole that reflects the communicative goal of the Resolution stage, to grasp and make sense of the happenings surrounding the conflict.

Genre in the classroom and across the curriculum

In the advanced-level class where 'Three Friends' is read, the story serves as the basis for one of the writing tasks. Students retell the story, but from the perspective of one of the other characters, and in so doing are expected to draw on a similar schematic structure to the one presented in the source text. A genre-based pedagogy that involves scaffolded support of modelling and text deconstruction leading towards students' independent construction of the text (Cope and Kalantzis 1993a) explicitly draws on the discourse and lexicogrammatical features outlined in this analysis. Students analyse the text according to its schematic structure and dominant lexicogrammatical features through individually assigned homework and in-class group and whole-group discussions in which students recount[3] and interpret the events. The instructional phases include:

(1) Identification of text breaks and major topics in 'Three Friends'. Attention is given to temporal expressions and conjunctions in signalling the beginning of the stages;
(2) Development of semantic fields represented across the text, i.e., friendship and estrangement, the army draft and the escape;

(3) Identification of voices constructed in the text with focus on the narrator's positioning role;
(4) Role-play activity, in which students take on one of the characters from the story and recount the events from their perspective;
(5) Introduction of the writing task;
(6) Written feedback on first drafts from the instructor before reworking final draft.

To make the role of genre and register explicit in the writing expectations across the curriculum, all task guidelines in the GUGD follow a tripartite task structure that addresses task appropriateness, content and language focus. *Task Appropriateness* outlines the specific situational context motivating the writing event, detailing the communicative purpose and writer and reader roles, as well as the generic structure of the expected text. *Content* refers to the writers' ability to fulfil the generic moves through engagement with the content material. Under *Language Use*, targeted language features at the discourse- and clause-levels deemed necessary for accomplishing the task are linked to corresponding generic stages. Assessment of first and second drafts of the tasks reflects the guidelines along each of the three task categories (see Byrnes 2002).

Placing genre and schematic structure in a central role in the writing process underscores the advantages of explicit instruction: learners have a step-by-step frame for fulfilling the expectations of the genre, which additionally strengthens their awareness of discourse-level features as they relate to communicative purpose, and instructors have an important resource for evaluating L2 learners' language use. As students progress through the curriculum and become accustomed to the genre-based writing task process, they are also likely to develop a better sense of the relationship between language and particular contexts of use.

An important offshoot of this writing approach within the GUGD curriculum has been acknowledgement by the faculty that understanding and fostering the understanding of textual models as representing stable, prototypical genres is fundamental for students' successful learning (Byrnes 2002). Educators therefore have the responsibility of determining which genres serve as effective models for targeted L2 language use as well as elicit thoughtful and useful engagement with content material. While prototypical texts are likely to be easier for beginning students (Flowerdew 2002), more varied, hybrid texts that integrate other textual patterns may be more suitable at advanced levels. Of course, modelled texts provide just one instantiation of how a genre may be realized and there are likely to be additional means by which a given genre can be construed linguistically (Eggins and Martin 1997). A pedagogy that stresses choice, therefore, would need to consider what such additional linguistic resources might look like and then integrate them, preferably in the form of texts, into instruction as appropriate.

Conclusion

I have argued that text and genre serve as useful curricular units for dealing with the challenge of charting the developmental trajectory of L2 learning from beginning to advanced levels, and that knowledge of the structural components of texts, in particular, can help curriculum designers and instructors map texts on to linguistic and knowledge building goals. Through attention to the recurrent structural properties of texts, the curriculum designer can see how linguistic forms and functional categories are realized in communicatively relevant ways. This has consequences for both materials development and pedagogical task design, constructs that in educational environments where text plays a key role must be mutually supportive of each other (for materials development issues, see Byrnes 2006).

Both the texts presented here serve as a basis for L2 student production. The recount used in the beginning-level classroom displays prototypical linguistic features for the genre, whereas the personal narrative used at the advanced-level draws on a variety of lexicogrammatical features to fulfil its generic stages, including incorporation of other genres. For AL2 reading involving complex texts that are comprised of hybrid genres, knowledge of how texts are constructed can help learners to uncover the symbolic significance of certain embedded genres. For AL2 writing, where a broadened linguistic repertoire means greater linguistic choice, nuanced knowledge of the larger meaning potential of registers and genres is essential for situating one's individual voice within recognizable communicative contexts.

As one observes the dominant genres represented in a curriculum's materials and tasks, the following questions can help highlight clearly the relationship between text, genre and task across a program:

(1) Are there any genres considered relevant to the teaching mission of a particular department that are absent across materials or tasks in the curriculum?
(2) What is the relationship between the genres of source texts and those of student tasks? Are students expected to produce the same genre that was modelled? Are register variables shifted or subverted (as in a parody)? Are the genres from the source text embedded in the task genre?
(3) Do source texts used for L2 production adequately highlight the genre's meaning potential? Do they represent prototypical genres, or variations of genres? Are optional stages also considered?
(4) Do students have a transparent model of how optional and obligatory stages are realized linguistically for the purposes of text comprehension or production? If not, do they need one?

These questions provide just a starting point for reflecting on curricula and considering the potential support that texts give each other across the curriculum. While I have focused primarily on the potential of narrativity, other written discourse types can appear across a curriculum and may play prominent roles in a department's curriculum. Considering the relationship between texts of similar communicative purposes and rhetorical structures should help the curriculum designer in devising pedagogical pathways to support the goal of attaining high AL2 abilities.

Notes

1 Here, narrativity refers to the larger rhetorical structure drawn on in such story genres within the SFL tradition as recounts, exemplums and personal narratives (see Plum 1988; Rothery and Stenglin 1997) as well as those genres representing the traditional literary category of prose. Though this paper approaches narrative from an SFL perspective, it is important to bear in mind the different traditions FL departments represent in using genre as a construct and consider how these traditions can best be negotiated given a program's specific educational goals.

2 The GUGD curriculum consists of five instructional levels: Level I to III each comprise two-semester sequenced courses, while Levels IV and V represent a variety of non-sequenced courses at advanced levels. See http://www3.georgetown.edu/departments/german/programs/curriculum/index-.html for more information on the GUGD undergraduate curriculum 'Developing Multiple Literacies'.

3 Byrnes and Sprang (2004) elaborate on the pedagogical support of 'Three Friends' as used in the GUGD curriculum, but from a cognitive processing standpoint. They illustrate how, with scaffolding, AL2 learners are able to develop the ability to narrate through use of discourse and lexicogrammatical features that create temporal and causal coherence.

References

Bakhtin, M. M. (1986) 'The problem of speech genres', in C. Emerson and M. Holquist (eds), *Speech Genres and other Late Essays*. Austin, TX: University of Texas Press, pp. 60–102.

Byrnes, H. (2001) 'Reconsidering graduate students' education as teachers: "It takes a department!" ' *The Modern Language Journal*, 85, 512–30.

Byrnes, H. (2002) 'The role of task and task-based assessment in a content-oriented collegiate FL curriculum'. *Language Testing*, 19, 419–37.

Byrnes, H. (2006) 'A semiotic perspective on culture and foreign language teaching: implications for collegiate materials development', in V. Galloway and B. Cothran (eds), *Language and Culture out of Bounds: Discipline-blurred Perspectives on the Foreign Language Classroom*. Boston, MA: Heinle Thomson, pp. 37–66.

Byrnes, H. and Maxim, H. M. (eds) (2004) *Advanced Foreign Language Learning: A Challenge to College Programs*. Boston, MA: Heinle Thomson.

Byrnes, H. and Sprang, K. A. (2004) 'Fostering advanced L2 literacy: a genre-based cognitive approach', in H. Byrnes and H. M. Maxim (eds), pp. 47–85.

Byrnes, H., Weger-Guntharp, H. and Sprang, K. A. (eds) (2006). *Educating for Advanced Foreign Language Capacities: Constructs, Curriculum, Instruction, Assessment*. Washington, DC: Georgetown University Press.

Christie, F. (2002) 'The development of abstraction in adolescence in subject English', in M. J. Schleppegrell and M. C. Colombi (eds), pp. 45–66.

Christie, F. and Martin J. R. (eds) (1997) *Genre and Institutions. Social Processes in the Workplace and School.* London: Cassell.

Coffin, C. (1997) 'Constructing and giving value to the past: an investigation into secondary school history', in F. Christie and J. R. Martin (eds), pp. 196–230.

Colombi, M. C. (2002) 'Academic language development in Latino students' writing in Spanish', in M. J. Schleppegrell and M. C. Colombi (eds), pp. 67–86.

Cope, B. and Kalantzis, M. (1993a) 'The power of literacy and the literacy of power', in B. Cope and M. Kalantzis (eds), pp. 63–81.

Cope, B. and Kalantzis, M. (eds) (1993b) *The Powers of Literacy: A Genre Approach to Teaching Writing.* Pittsburgh, PA: University of Pittsburgh Press.

Eggins, S. and Martin, J. R. (1997) 'Genres and registers of discourse', in T. A. van Dijk (ed.), pp. 230–56.

Eigler, F. (2001) 'Designing a third-year German course'. *Die Unterrichtspraxis*, 34, 107–18.

Flowerdew, J. (2002) 'Genre in the classroom: a linguistic approach', in A. M. Johns (ed.), pp. 91–102.

Gee, J. P. (1998) 'What is literacy?', in V. Zamel and R. Spack (eds), *Negotiating Academic Literacies: Teaching and Learning across Languages and Cultures.* Mahwah, NJ: Lawrence Erlbaum, pp. 51–9.

Halliday, M. A. K. (1985) *An Introduction to Functional Grammar.* London: Edward Arnold.

Halliday, M. A. K. (1999) 'The notion of "context" in language education', in M. Ghadessy (ed.), *Text and Context in Functional Linguistics.* Amsterdam/Philadelphia: John Benjamins, pp. 1–24.

Hasan, R. (1984/1996) 'The nursery tale as a genre', in C. Cloran, D. Butt and G. Williams (eds), *Ways of Saying: Ways of Meaning. Selected Papers of Ruqaiya Hasan.* New York: Cassell, pp. 51–72.

Hieber, W. (1998–2005) 'Am Brandenburger Tor in Berlin'. http://www.kaleidos.de/alltag/rituale/silv03.htm.

Johns, A. M. (ed.) (2002) *Genre in the Classroom.* Mahwah, NJ: Lawrence Erlbaum.

Klecker, J. (1991) 'Drei Freunde', in K. Humann (ed.), *Denk ich an Deutschland – Menschen erzählen von ihren Hoffnungen und Ängsten.* Hamburg: Luchterhand.

Kress, G. (1993) 'Genre as social process', in B. Cope and M. Kalantzis (eds), pp. 22–37.

Labov, W. and Waletzky, J. (1967) 'Narrative analysis: oral versions of personal experience', in J. Helm (ed.), *Essays on the Verbal and Visual Arts: Proceedings of the 1996 Annual Spring Meeting of the American Ethnological Society.* Seattle: University of Washington Press, pp. 12–44.

Macken, M. R., Martin, J. R., Kalantzis, M., Rothery, J. and Cope, B. (1989) *An Approach to Writing K-12: The Theory and Practice of Genre-based Writing, Years 3–6.* Sydney: Literacy and Education Research Network in conjunction with NSW Department of Education, Directorate of Studies.

Macken-Horarik, M. (2002) '"Something to shoot for": a systemic functional approach to teaching genre in secondary school science', in A. M. Johns (ed.), pp. 17–42.

Martin, J. R. (1984) 'Language, register, and genre', in F. Christie (ed.), *Children Writing: Reader.* Geelong, Australia: Deakin University, pp. 21–9.

Martin, J. R. (1985) 'Process and text: two aspects of human semiosis', in J. D. Benson and W. S. Greaves (eds), *Systemic Perspectives on Discourse. Selected Theoretical Papers from the 9th International Systemic Workshop.* Norwood, NJ: Ablex, pp. 248–74.

Martin, J. R. (1992) *English Text: System and Structure.* Amsterdam/Philadelphia: John Benjamins.
Martin, J. R. (2002a) 'A universe of meaning – How many practices?', in A. M. Johns (ed.), pp. 269–78.
Martin, J. R. (2002b) 'Writing history: Construing time and value in discourses of the past', in M. J. Schleppegrell and M. C. Colombi (eds), pp. 87–118.
Martin, J. R. (2003) 'Making history. Grammar for interpretation', in J. R. Martin and R. Wodak (eds), *Re/reading the Past: Critical and Functional Perspectives on Time and Value.* Amsterdam/Philadelphia: John Benjamins, pp. 19–57.
Martin, J. R. and Plum, G. (1997) 'Construing experience: Some story genres'. *Journal of Narrative and Life History*, 7, 299–308.
Matthiessen, C. M. I. M. (1993) 'Register in the round: diversity in a unified theory of register analysis', in M. Ghadessy (ed.), *Register Analysis: Theory and Practice.* London: Pinter, pp. 221–92.
Ochs, E. (1997) 'Narrative', in T. A. van Dijk (ed.), pp. 185–207.
Plum, G. A. (1988) 'Text and Contextual Conditioning in Spoken English: A Genre-based Approach'. Unpublished doctoral dissertation, University of Sydney.
Rothery, J. and Stenglin, M. (1997) 'Entertaining and instructing: exploring experience through story', in F. Christie and J. R. Martin (eds), pp. 231–63.
Rothery, J. and Stenglin, M. (2000) 'Interpreting literature: the role of appraisal', in L. Unsworth (ed.), *Researching Language in Schools and Communities: A Functional Linguistic Perspective.* London: Cassell, pp. 222–44.
Schleppegrell, M. J. (2002) 'Challenges of the science register for ESL students: errors and meaning-making', in M. J. Schleppegrell and M. C. Colombi (eds), pp. 119–42.
Schleppegrell, M. J. and Colombi, M. C. (eds) (2002) *Developing Advanced Literacy in First and Second Languages. Meaning with Power.* Mahwah, NJ: Lawrence Erlbaum.
Swaffar, J. and Arens, K. (2005) *Remapping the Foreign Language Curriculum: An Approach through Multiple Literacies.* New York: MLA.
Swales, J. M. (1990) *Genre Analysis: English in Academic and Research Settings.* Cambridge: Cambridge University Press.
Van Dijk, T. A. (ed.) (1997) *Discourse as Structure and Process. Discourse Studies: A Multi-disciplinary Introduction.* Volume I. London: Sage.
White, P. (1997) 'Death, disruption and the moral order: the narrative impulse in mass-media "hard news" reporting', in F. Christie and J. R. Martin (eds), pp. 101–33.

12 Advanced language for intermediate learners: corpus and register analysis for curriculum specification in English for Academic Purposes[1]

Nick Moore

Introduction

While communication skills in English are essential for students who are enrolled in English for Academic Purposes (EAP) around the world, it is the academic, professional and vocational skills necessary to obtain suitable employment after graduation that remain their primary goal. That functional or practical orientation is both a challenge and an opportunity for students, academic institutions and language teaching programs that use English as the language of instruction outside English-speaking countries.

In part, the challenge derives from political and economic pressures demanding efficiency and effectiveness from EAP programs to enable their students to attain career goals. The pressure is on EAP programs to accelerate learners towards their academic careers, where they need to cope with advanced language, even though most of them would only qualify as intermediate learners according to international standardized tests, such as TOEFL or IELTS, the International English Language Testing System. Beyond simple demands for efficiency, however, the challenge also lies in programs having to give careful consideration to the relation between content, particularly disciplinary content, and language use, a relation that is typically viewed from a pedagogical perspective (e.g., Brinton *et al.* 1989) rather than in terms of its linguistic implications (cf. Mohan 1986; Schleppegrell 2001; Schleppegrell *et al.* 2004).

EAP learners assist in meeting this challenge by bringing to the learning task the motivation, taxonomies and schemata related to studying a discipline, and the experience of learning an L1. With this profile, EAP learners question the notion that all learners must learn the whole of the language. Increasingly, a sociocultural understanding of language has foregrounded the realization that becoming integrated into a professional community is fundamentally a language-based process, a form of socialization into the discourse community of a discipline or profession. In that sense, EAP programs have an opportunity to rethink some of the field's foundational assumptions, particularly as they relate to definitions of advanced language and goals of language use.

This paper is a contribution to such deliberations. It recommends placing *register* at the centre of program building and pedagogical recommendations because of the explicit link that register recognizes between variation in contexts of use, including academic and professional contexts, and variation in the occurrence of linguistic features. From among the issues that arise within a register approach (see Matthiessen 1993), this paper highlights the special role of lexicogrammatical features. It proposes that corpus-based analyses allow for detailed specification of those lexicogrammatical items that, due to a higher frequency of occurrence than would be expected when compared to language use in the system as a whole, distinguish one register from another. Assuming academic language to be the target acquisitional goal of an EAP program, such an approach will distinguish both features specific to a register – in this case, academic (sub-)disciplines – and general features of the language system with the 'highest surrender value' from those features that are rarely, if ever, used in a specific register. In the identification of relevant lexicogrammatical features, corpus-based analysis can make important contributions to curriculum construction, pedagogies and assessment practices in EAP environments, particularly when it retains a textual and sociocultural orientation in its design, analysis and interpretation.

A register orientation immediately exposes the limited value to EAP of the verb-centred paradigm of prevailing structure-based curricula. It also queries some central assumptions in language learning by literate adults. Do all learners need to learn all structures? Is it possible for adult learners to circumvent linguistic features that are infrequent in the target register? For instance, can instructed adult learners move relatively quickly from the congruent semiosis of verb-centred ways of construing processes to the non-congruent nominalization of processes that populate English academic texts? Halliday (also Halliday and Martin 1993; Halliday and Matthiessen 2004) refers to this feature as grammatical metaphor and considers it a critical consequence of academic literacy (for related discussions see the contributions by Colombi, Ryshina-Pankova and Schleppegrell, this volume). Thus, this paper contributes to a rethinking both of notions of 'advancedness' and of the L2 pedagogies that are most suited for literate adult learners.

A new description of language in ESL and EFL

A survey of English language teaching textbooks, pedagogical grammars and teaching resources reveals that great emphasis is consistently placed on the verbal group, with tenses, aspects and functions using modal auxiliaries providing both the content and the sequencing of instruction. Thus, present tenses and aspects tend to precede past tense which precede future time. Simple tenses precede continuous forms, which precede the perfect aspect, all of which precede combinations of verb forms. In this paradigm,

'Advanced language' is synonymous with the 'future perfect', 'mixed conditionals' and other complex verbal group structures.

Verb-based curriculum sequencing, which appears to be partly derived from morphology acquisition studies and the 'natural order hypothesis' (Dulay and Burt 1974) popularized by Krashen (e.g., 1985), is generally accepted within EFL curricular recommendations, pedagogies and materials. Even so, reservations regarding the notion that all second language learners 'acquire' linguistic structures in a particular sequence do exist (e.g., Larsen-Freeman 1975). Of the various directions that challenge has taken, four are highlighted here.

Most fundamentally, the notion 'language acquisition' itself has been variously criticized, especially when it is limited to a narrow range of linguistic structures:

> The implication has been that the learning of structure is really at the heart of the language learning process. And it is perhaps not too far-fetched to recognise in the use of the term *acquisition*, a further implication that structure, and therefore language itself, is a commodity of some kind that the child has to gain possession of in the course of maturation. (Halliday 1975: 1, emphasis in original)

This objection focuses on the model of learning-as-acquisition as failing to recognize language learners' essential capacity to actively create their own meanings, and as considering learning to be nothing more than acquiring a limited, and largely undefined, model. Furthermore, critical perspectives question such positivist notions of linguistic competency, proposing that the practice of (structure-based) second language teaching and learning may construct its own self-serving circular definitions of competency (McNamara 2001).

Second, register theory as described below questions the concept of language as a single construct, and, more specifically, as primarily a grammatical construct. Instead, it foregrounds the centrality of variation and choice of lexicogrammatical resources within a social context.

Third, a reorientation to sociocultural and textual definitions of the target language as is implied in a focus on registers recognizes that considerable exposure to the target language is necessary before such language use can be integrated into an individual's meaning potential. But that recognition stands quite apart from another insight, namely that the meaning potential available to an adult differs significantly from that of a first language (L1) learner. While Vygotsky (1966/1991) focuses on the socio-psychological processes involved in mental maturation, Halliday (1993) details a 'natural order' of learning that is based on linguistic development, following the sequence of *protolanguage*, *generalization*, *abstractness* and *metaphor*. Assuming that adults' L1 linguistic and mental development has progressed to the final stages of *grammatical metaphor* and *synoptic/dynamic complementarity* (Halliday 1993), an adult L2 learner, unlike a young L1 or L2 learner, can benefit from a program that reflects the distribution of lexicogrammatical features of a target register, even if the features are mostly grammatical metaphor and therefore presumed to be late-acquired. This results in an efficient language-learning program that enables adult learners, who may not be able to function in all L2 registers or genres, to gain access to the meaning-making resources most commonly associated with a professional discourse community. They become 'intermediate' language learners functioning effectively in advanced language.

Finally, corpus studies provide empirical evidence to challenge the notion that fixed grammatical structure is at the heart of language in use and, therefore, the primary criterion for planning language learning programs. Among other issues to be considered, for example, von Stutterheim and Carroll (2006) detail the considerable influence of L1 lexicogrammatical categories on the construal of experience even in the language use of advanced L2 learners.

A theory of register

If one abandons a focus on generalizable and fixed grammatical structures as being the central characteristic of language, variation becomes a paramount feature of language use. Language variation is a product of variation in the context of situation within the context of culture. The context of culture is defined by social variables such as social hierarchies, region and historical moment. Language variation in the context of culture produces socially sanctioned patterns of behaviour, including verbal behaviour. As language is repeatedly called upon to produce the same results, its use becomes conventionalized. These conventional patterns constitute *genre*. Genre tends to stage, with both required and optional steps, sequences of activities that produce socially acceptable goals (Halliday and Hasan 1985; Ventola 1987). The relationship between genre and register is one of realization; a genre is realized in the context of situation, or register (Leckie-Tarry 1995; Martin 1984/2001, 1992, 1997, 1999).

Language in the context of situation co-varies with the configuration of Field, Tenor and Mode. Register is an attempt to characterize configurations within a speech community that arise to limit the textual choices a speaker will make in a particular context from among the options of the language system as a whole. To date, the most extensive analysis of the situational constraints on language in use has been developed by Halliday (e.g., 1978), Hasan (e.g., 1979) and other systemic functional linguists (e.g., Matthiessen 1993). Generally, within Systemic-Functional Linguistics (SFL), Field can be thought of as the subject under discussion, Tenor as the relationship between those involved in the text and Mode as the way the message is being delivered. These terms are not arbitrary, but align themselves with the three metafunctions of language. Field is typically realized in the Ideational metafunction, which reflects our material reality, Tenor in the Interpersonal metafunction, which enacts exchanges, and Mode in the Textual metafunction, which creates messages (Halliday and Hasan 1985; Halliday and Matthiessen 2004). These three metafunctions encompass the meanings of language structures, extending from phonology to discourse structure (Martin 1992). Table 12.1 illustrates the typical correspondence between register and structure.

The relationship between register and linguistic form can be readily observed in the kinds of lexicogrammatical choices that realize the Field in a particular context. Each register contains combinations of indexical lexical items as well as typical grammatical patterns that identify the Field. Academic text, for instance, is characterized by a large proportion of relational processes, among them relationships such as definitions and properties (Halliday and Martin 1993; Martin 1989; Martin and Veel 1998). Reflecting the systemic nature of this approach to language analysis, it is the combination of lexicogrammatical features, not the features themselves, that reveal the field. In the following extract, indexical lexical items are highlighted in **bold**, while relational processes are in *italics*.

> **The Binary Number System (B)**
> The internal mechanisms of most **digital computers** *involve* **devices** that flip back and forth between two **states**. The **relays** of the previous section *are* either **off** or **on**, the **electrical** wires either carry **current** or do not, and so forth. Since the **components** of the **computer** *are* only two-valued, a reasonable way to *represent* **states** *is* with two-valued notation, which *is* why **information** *is* **coded** into **binary** form. (Biermann 1997: 240, all emphases added)

Interpersonal choices, including choices in mood and modality that reflect the degree of commitment and solidarity, constitute the Tenor of a text, the second dimension of register. The relationships enacted in text through linguistic choices reveal the status of the participants. In academic textbooks this is often the unequal status of expert author and novice reader. Modality in academic text is revealed in a range of 'hedging' devices, such as *generally*, *tendency*, *claim*, and through conventions of citation and acknowledgement that indicate the writer's position vis-à-vis the statements made (Hyland 1998, 1999, 2002b).

Table 12.1 Meaning of Field, Tenor and Mode, alignment with metafunctions, and typical realizations

Register Variable	'*Mnemonic*': Function in Language	Metafunction	Typical Realizations
Field	*Topic* reflects material reality	Ideational	Combinations of indexical lexical items; typical process types
Tenor	*Relationship* enacts exchanges	Interpersonal	Mood, Modality
Mode	*Role of Language* creates messages	Textual	Deictic devices, Theme, Information structure

Finally, the register of a text is revealed in the textual metafunction. For instance, because face-to-face verbal interaction can draw on a shared physical context, deictic devices become highly favoured textual resources, whereas written text needs to establish its own context through text (Halliday 1989). Thus, a sentence like 'Pass me that brown one over there' relies almost entirely on a shared physical context, and seems to convey little ideational meaning. By contrast, written academic texts frequently refer to themselves by asking questions, making predictions and negotiating with the reader (Hyland 2002a; Tadros 1989; Thompson 2001): that is, they draw on resources that instantiate the textual metafunction. In sum, a theory of register assists in explaining how academic text has its own patterns of language which can then be identified and included in a language program.

The probabilistic nature of register

SFL highlights meaning-making as the product of (not necessarily conscious) choices made within systems in each metafunction, rather than adherence to fixed rules and structures. Each choice derives its meaning from contrast to the option(s) not selected within a certain context. In the interpersonal metafunction, for instance, the choice of Mood, between an indicative or imperative clause, or between a Wh-question for subject, complement or adjunct (see Fig.12.1), creates meaning.

Contributing to the meaning of each choice is the likelihood of its selection in a specific environment, a consideration that creates the concept of 'unmarked' for most probable and 'marked' for least probable occurrence when likelihood of selection is skewed. Halliday (1956/1976) established this principle in Chinese using a manual count of selected systems and extended it by examining English primary tense and polarity. Halliday and James' (1993) corpus study supported the findings of Nesbitt and Plum (1988) that systems tend to be heavily skewed (90:10), skewed (70:30) or equal (50:50). An example of Nesbitt and Plum's findings is presented in Figure 12.2 below, which focuses on the preference in English for direct speech and reported ideas.

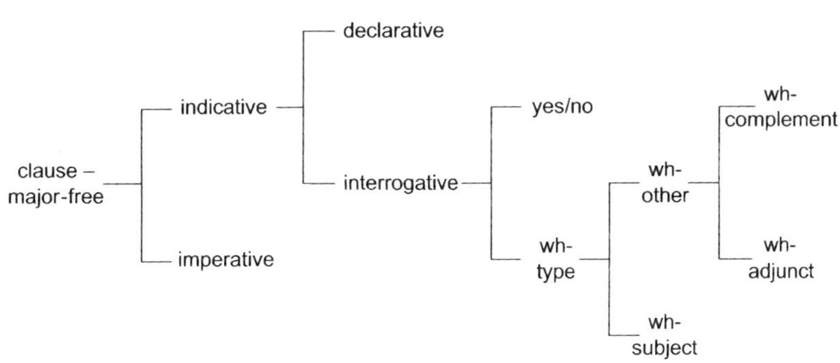

Figure 12.1 Extract of Mood system

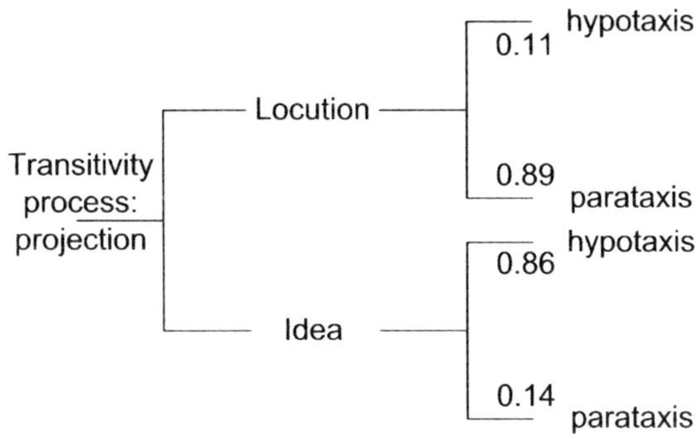

Figure 12.2 Probabilities of clause types in Locution and Idea process types

The very definition of register can be connected to the probability of features being chosen. Halliday (1991a) points out that we recognize a register because it makes meaningful selections with probabilities that do not match the system as a whole. Work is being carried out (Matthiessen, forthcoming) to increase the number of contexts and subsystems of language for which we have the probabilities of selection. None of these variations represents a new language, however. Instead, each variation represents a change in the likelihood of making choices within the systemic network; each variation represents a register.

Importantly for EAP programs, the probability of selecting from the choices offered by the context of situation within the context of culture affects grammar and lexis equally, motivating the preference in SFL for the term *lexicogrammar*:

> In fact lexis and grammar are not different phenomena; they are the same phenomenon looked at from different ends. . . . it is the probabilistic model of lexicogrammar that enables us to explain register variation. Register variation can be defined as the skewing of (some of) these overall probabilities, in the environment of some specific configuration of field, tenor and mode. It is variation in the tendency to select certain meanings rather than others. (Halliday 1991a: 57)

That is, all lexicogrammatical features contribute to realizing register variation. As such, structures that have thus far been considered as 'advanced' by the 'natural order hypothesis' and have been used to sequence a conventional EFL curriculum can now be reinterpreted as occurring with higher or lower frequency in different registers, and therefore as more or less relevant to the needs of particular groups of learners. Register theory, then, not only questions whether advanced language capacity can be considered a single construct, it also foregrounds the context-dependent nature of language use.

A theory of register allows us to recognize the language variation produced in the range of contextual configurations that a student is likely to encounter. To facilitate the next step, namely to ensure maximum effectiveness for an EAP program, the most relevant and useful lexicogrammatical features of academic registers need first to be identified and then sequenced within a syllabus. Both large- and small-scale corpus linguistics projects can provide results detailed enough to specify language for effective EAP programs.

Corpus evidence for register in EAP

Corpus analysis has revealed a number of important findings pertinent to EAP. Drawing on a multi-million word corpus, Biber *et al.* (1999) demonstrate that the use of the 'progressive' or 'continuous' aspect is very rare in academic text (Fig. 12.3). For all genres, the vast majority of simple aspect verbal groups contrast with a small number in perfect aspect, while conversation and fiction select perfect and progressive aspects far more frequently

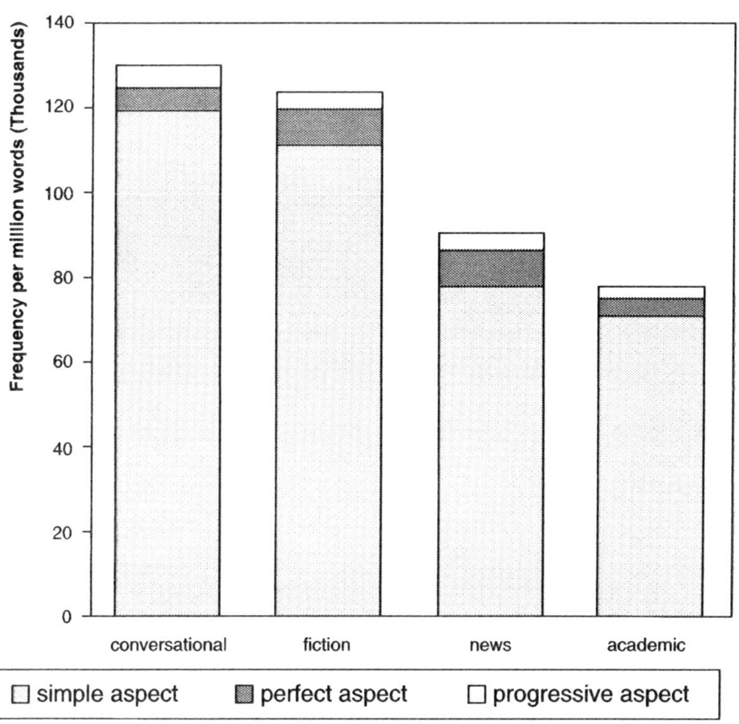

Figure 12.3 Frequency of simple, perfect and progressive aspects across text types (from Biber *et al.* 1999: 461)

than does academic text. Figure 12.4 (below), another example from Biber *et al.* (1999), reveals an even greater contrast with regard to noun modification, with pre- and post-modification occurring very rarely in conversation. By contrast, it is normal for a noun to be modified in academic texts, with a major proportion of nominal groups containing both pre- and post-modifiers. That is, in academic text the majority of meanings are packed into the nominal group (Halliday and Martin 1993; Martin 1989; Ventola 1996), unlike conversational texts, which locate many meanings in conjunctive clause structures.

This reality of academic English contrasts sharply with the vast majority of ESL and EFL course books, pedagogical grammars and language-teaching resources, which focus heavily on verb tenses and aspects, with little emphasis on the construction of noun groups or their modification. It appears that the emphasis on oral interaction in the 'communicative approach' has skewed the selection of structures in curricula to the advantage of conversational English and to the detriment of academic registers. Furthermore, the findings of Biber *et al.* (1999) illustrate clearly the point made by Halliday about

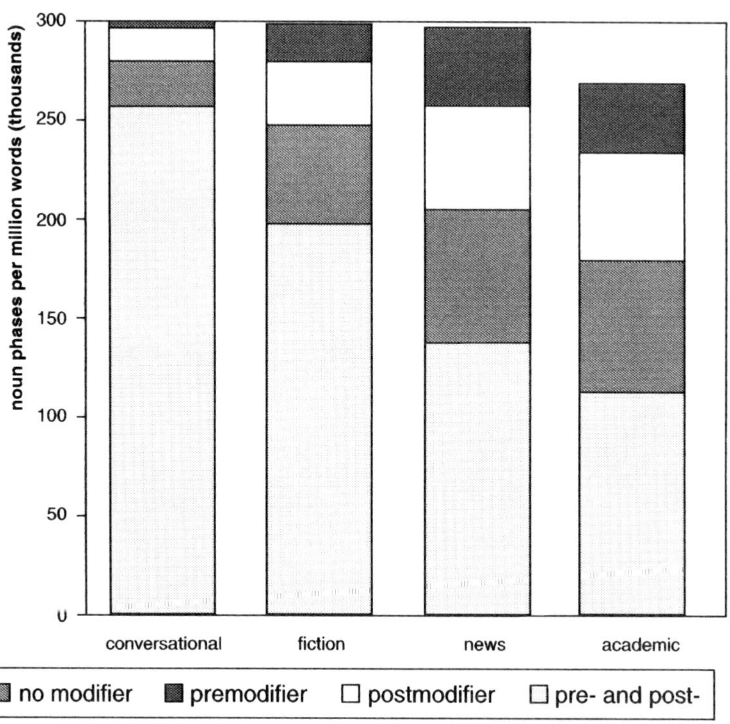

Figure 12.4 Frequency of modified and unmodified noun phrases, with type of modification, across text types (from Biber *et al.* 1999: 578)

the nature of register variation: from one register to the next, none of the potential choices in the language system change. What changes is the probability of choosing one form over another.

The findings of large-scale corpus projects are intended to be as generalizable as possible. Thus, Biber *et al.* (1999) make observations about 'conversational English' or 'academic text' as a whole. However, register is a gradable concept and can be identified to ever-greater levels of detail, or 'delicacy'. Each local context will reveal its own patterns. While large-scale corpora (e.g., the British National Corpus (BNC), http://www.natcorp.ox.ac.uk/) contain hundreds of millions of words to represent the language in its entirety, and medium-scale corpora aim to represent and analyse one variety of English (e.g., Michigan Corpus of Academic Spoken English, http://www.hti.umich.edu/m/micase), smaller corpora can prove invaluable in identifying relevant language for a specific group of students (Flowerdew 1993; Ghadessy *et al.* 2001). A corpus developed in the local context will accurately identify the lexicogrammatical features needed most frequently by students in particular disciplines and sub-disciplines within individual institutions.

We can look at this exercise as a form of needs analysis, one that examines the language required by students. By building a local corpus to represent the target language of a group of students, curriculum planners gain an indispensable tool for identifying relevant features of the lexis, structure, discourse and genres of academic texts. The rapid growth in computer processing and storage power and the increasing digitization of text and images has put corpus linguistics within reach of most language teachers with the minimum of computer resources. An analysis of the language required by a particular group of students can be performed relatively easily and cost-effectively by building a local corpus. After the necessary copyright permissions have been gained and the security of the data has been assured, digital text can be easily stored for analysis. With voice recognition software becoming increasingly accurate, it is also becoming more cost-effective to develop machine-readable samples of the spoken academic text that is produced at any institution. (Incidentally, questions of copyright permission, confidentiality and security are no less important here than for published materials.)

The aim of a local corpus project is to examine specific registers. To reveal differences in frequency between a register-specific corpus and a general, or 'reference' corpus, we can use the following formula (Fig. 12.5) for the 'Register Variance Differential (RVD)', (based on the Aston Text Analyser 1.0 (Roe 1995) and Dunning 1993):

$$10^4\left(\frac{f_o}{s_o}\Big/\frac{f_e}{s_e}\right) + \log\left(10^4\left|\frac{f_o}{s_o} - \frac{f_e}{s_e}\right|\right)$$

Figure 12.5 Formula for Register Variance Differential (RVD)

f_o is the observed frequency of an item in the corpus under study, f_e is the expected frequency of that same item, s_o the total number of tokens in the observed corpus, and s_e the total tokens in the reference corpus.

The main task of the RVD is to accentuate the difference in expected and observed relative frequency scores of a lexical item between a reference corpus and the register-specific corpus. That is, it quantifies the variation in linguistic choices for a particular register in comparison with the language system as a whole. The expected score from the reference corpus must therefore come from a reliable source, such as Cobuild or the BNC, designed to include as wide a variety of registers as possible, so that no single register can influence the frequency count of any lexicogrammatical item.

The RVD compares the number of times a word is used in general (corpora) with the number of times it is used in a register (-specific corpus), per 10,000 words in this instance, and accentuates the difference between the two scores. As the RVD is a measure of the variation in relative frequency, it tells us whether a word is *probably* being used in a way that differs from what we would normally expect. For pedagogic purposes, the RVD scores can be divided into four groups called *General*, *Specialized*, *Sub-Technical* and *Technical* as described and set out in Table 12.2, which outlines a vocabulary syllabus with examples from an approximately 225,000-word, highly register-specific corpus from the field of software engineering.

Table 12.2 Definition, description and examples of categories of Register Variance Differential (RVD)

Category	Technical definition	Description	Examples
General	RVD ≤ 4	Little or no variation in frequency between specialist and general corpora	the, moved, speed, of, answer, separate, primary
Specialist	4< RVD <10	Slight variation in frequency between corpora	assist, eliminate, illustration, requests, organization
Sub-technical	RVD ≥ 10	Substantial difference in frequency between corpora	software, x, file, code, development, value, memory
Technical	No RVD: Frequency < 1 in 1,000,000 in reference corpus	Frequent in specialist corpus, but less than once in 1 million words in 'general' English	cpu, compiler, dialog, scanf, Linux, algorithm

General vocabulary items exhibit frequency patterns in the corpus under study that are similar to the reference corpus and so are probably being used in the same way across a wide range of registers. This group includes both high and low absolute frequency items. To give students the best return for their vocabulary-learning effort (Coxhead and Nation 2001), they need not learn low frequency items from any category. *Specialist* terms show a small variation in the register-specific corpus, which suggests these words are only found in a narrow range of registers. For instance, they may be items associated with journalism, technical manuals or academic prose. *Sub-technical* words exhibit a large variation in relative frequency between a general purpose corpus and a specific purpose corpus. These items are found with far lower frequency in a general corpus, suggesting that the item has a specialized pattern of usage in the corpus, and so a specialized meaning. Finally, the category of *technical* lexis assembles indexical words that are relatively frequent in the specialist corpus but non-existent or extremely rare (i.e., occurring less than once in a million words) in the general corpus. The four groups, and their cut-off points, are arbitrary as far as the RVD score is concerned, but reflect the nature of lexicogrammatical items and the choices available to EAP syllabus designers.

Placing lexis at the heart of a syllabus for EAP

Corpus studies of the English language have revealed that a very small proportion of words account for a very large proportion of text; every corpus contains a small number of high frequency words, and a large number (up to 50%) of low frequency words. No matter the size of the corpus or the language variety, the pattern represented in Figure 12.6, which shows a logarithmic scale for frequency per 10,000 words in a corpus on the y-axis and the number of items along the x-axis, repeats itself. The graph represents an analysis of a 200,000-word mixed-register opportunistic corpus of texts freely available on the Internet.

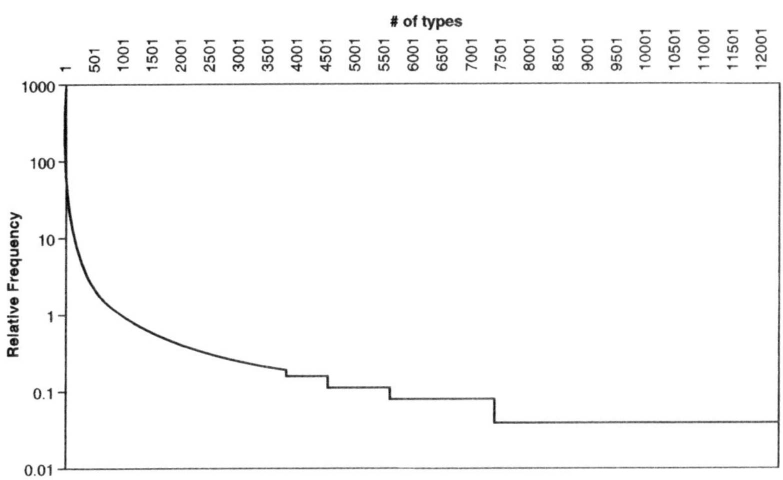

Figure 12.6 Distribution of the relative frequency of lexical items in a corpus

Across all registers, approximately 2,000 words are used to produce about 80% of English text (Sinclair and Renouf 1988; Willis 1990). The most common 100 words in English (itemized in Table 12.3) are generally known as 'grammar' or 'function' words, and even the most frequent nouns (#70 *time*, #72 *people* and #90 *way*) often function anaphorically. More importantly for a theory of grammatical structure, these frequent words constantly combine and recombine. Thus, the overwhelming statistical evidence for the repetition of sentence frames, set phrases, collocations, fixed combinations (Nattinger and DeCarrico 1992; Wray 2002) and other consistent patterns such as lexical priming (Hoey 2004, 2005) disputes a model of language that takes grammar, 'filled' by appropriate lexis, as the basic building block of language: this model is unable to account for the fact that we repeatedly combine the same lexicogrammatical features. Despite the hundreds of thousands of words in the English language, fewer than 5,000 word forms account for about 90% of even the most sophisticated text (Ward 1999). There is, then, a 'core' vocabulary (Carter and McCarthy 1988) that will probably be found in all registers.

Highlighting fixed syntagmatic patterning in language in no way denies creativity in language. Innovative combinations of words and meanings exist, most noticeably in poetry and highly valued prose, precisely because they stand in contrast to typical combinations of language; they are, literally, the exception to the norm. Sinclair (1991) coined the term 'The Open Choice

Table 12.3 First 100 word forms in the Birmingham Corpus, ranked in order of frequency of occurrence (from Sinclair and Renouf 1988)

1 the	21 they	41 what	61 has	81 because
2 of	22 at	42 their	62 him	82 two
3 and	23 his	43 if	63 some	83 over
4 to	24 have	44 would	64 into	84 don't
5 a	25 not	45 about	65 then	85 get
6 in	26 this	46 no	66 now	86 see
7 that	27 are	47 said	67 think	87 any
8 I	28 or	48 up	68 well	88 much
9 it	29 by	49 when	69 know	89 these
10 was	30 we	50 been	70 time	90 way
11 is	31 she	51 out	71 could	91 how
12 he	32 from	52 them	72 people	92 down
13 for	33 one	53 do	73 its	93 even
14 you	34 all	54 my	74 other	94 first
15 on	35 there	55 more	75 only	95 did
16 with	36 her	56 who	76 it's	96 back
17 as	37 were	57 me	77 will	97 got
18 be	38 which	58 like	78 than	98 our
19 had	39 an	59 very	79 yes	99 new
20 but	40 so	60 can	80 just	100 go

Principle' to describe how language users combine words into original combinations in a 'slot-and-filler' model, which takes much greater cognitive effort and is therefore less frequent. The processes involved for advanced L2 learners in employing the open choice principle have been illuminated by research into the considerable role played by L1 grammatical categories in the structuring of information (Carroll and Lambert 2006; von Stutterheim and Carroll 2006). 'The Idiom Principle' (Sinclair 1991), by contrast, describes language use derived from pre-fabricated units; based on corpus evidence, Sinclair hypothesized that it is the *modus operandi* of language users. For instance, when we say '*Yes, but the thing is . . .*' the phrase is probably not generated constituent-by-constituent, but is most likely available for use, 'pre-packaged' with its own meaning. The Cobuild project suggests that the majority of language shares this feature. Novel, innovative language is less likely to be employed in spontaneous speech because of real time constraints (Pawley and Syder 1983), but can commonly be found alongside frequent lexicogrammatical patterns in reworked texts. This radical revision of the relationship between lexis and grammar demands that lexis play a key role in representing the language typical of a particular register, and therefore will contribute greatly to a reliable EAP syllabus.

The results of corpus projects, particularly COBUILD (Sinclair 1987), have led to various proposals to replace structure with lexis as the main organizing principle in a syllabus (Lewis 1993; Nattinger and DeCarrico 1992; Sinclair 1991; Willis 1990). Corpus results, such as RVD scores, from well-designed corpora provide specification of lexical items which, when used as the main sequencing criterion in a curriculum, necessarily introduce all common structural patterns in proportion to their frequency because the most common lexical items (re- and co-) occur in common structures (Willis 1990, 2003). That is, there is no empirical division between lexis and grammar.

Table 12.4 reveals why an accurate lexicogrammatical syllabus is so vital to a successful academic career. Using results from a register-specific corpus for software engineering students, the final column represents how many unknown words the student is likely to come across in technical text at each level of vocabulary development. No matter how advanced an elementary student's skills in inferring meaning from context, knowing only one in every 2.6 words in a sentence in unsimplified text will not provide enough context for guesswork of an unknown item to take place (Hirsh and Nation 1992; Paribakht and Wesche 2006). It is only with the combination of all four categories, totalling about 4,500 types or 2,500 'word families' (similar to lemmas (Bauer and Nation 1993)), that students will be confident in their ability to cope with the advanced language in technical text because they will know an average of 12.5 words in the context of one unknown term (Nation 1999; Nation and Waring 1997). That is, a student can deal with advanced language with a vocabulary of approximately 2,500 lemmas (Coxhead 2000; Moore 2000; Nation 1999), or between 3,000 (Willis 1990) and 5,000 (Ward 1999) lexical types.

Table 12.4 Summary of vocabulary syllabus and effect on reading ability

To be taught in . . .	Category	# Types	# Word families (approx.)	# Tokens	Cumulative % of text	1 Unknown word in . . .
Foundation program	General	1863	790	139,022	61.8	2.6
Pre-sessional course	Specialist	639	278	22,151	71.6	3.5
In-sessional support	Sub-technical	684	334	34,487	87	7.7
Subject course	Technical	1328	1114	12,639	92.6	13.5
	Total	4514	2516	208,299		

Combining the lexical results of the local corpus with those patterns identified in large-scale corpora and confirmed in the register-specific corpus of academic texts, it is possible to specify in great detail the stages of an EAP language learning syllabus that will enable an adult language learner to be exposed to, notice and proceduralize (Schmidt 1990, 1993) the language that is typical of a discipline, including the most common combinations and collocations of frequent lexicogrammatical items. Language specified by the syllabus can be checked during text selection to ensure that the register is accurately represented, during materials design to select the most suitable lexicogrammatical patterns to be learned, and after the program to check progress against the target language. Most of these steps can be automated using software such as VocabProfile (Nation 2002) or WordSmith Tools (Scott 2005). The lexicogrammatical syllabus can also be used to assess students, placing them at different levels in the curriculum or indicating different standards of attainment.

The role of corpora and register in specifying an EAP curriculum

I have argued for a register-based analysis as the foremost method to capture the essential qualities of English for Academic Purposes and for an understanding of register-based language use as being essentially lexicogrammatical in nature. In so doing, I have questioned the earlier hypothesis of a natural order of learning for structures in a second language (L2) on the basis of a sociocultural and textual orientation. That argument is corroborated by both carefully conceived corpus-based analyses that also provide a wealth of support for a non-structure based approach, and by studies of language acquisition (Ellis and Sinclair 1996; Painter 1989; Peters 1983).

Just as a dialect or idiolect predicts that certain lexicogrammatical choices will be favoured or avoided, registers, by definition, will tend 'to select certain combinations of meaning with certain frequencies' (Halliday 1991b: 33). That is, both lexis and structure will be used with varied frequency in different registers. The results of large-scale corpus studies and discourse analyses suggest that academic text is generally characterized by simple tenses (Biber *et al.* 1999), relational processes of Identity or Attribute, and by packing huge amounts of information into the noun phrase (Halliday and Martin 1993; Ventola 1996; also Colombi, Ryshina-Pankova and Schleppegrell, this volume). Locally constructed corpora can verify, exemplify and make explicit these findings, while also identifying the items that fall into the categories of *General*, *Specialist Sub-Technical* and *Technical* lexis in order to sequence learning. Using lexis to specify a language-learning program does not exclude structure. On the contrary, the most important grammatical patterns represented in a register are revealed by corpus studies. EAP students require language programs that offer opportunities to focus on these aspects of language. Anchoring a curriculum in corpus-based findings of academic registers can lead to an EAP program that enables 'intermediate-level' students to function in the advanced language of a discourse community.

Note

1 I would like to thank Profs Matthiessen and Vernon and my colleagues Mrs O'Brien and Mrs Burns for guiding me towards clearer explanations, and for their support and enthusiasm. I am also indebted to the editor, Heidi Byrnes, for her careful comments and considered suggestions. Despite this guidance, faults may still persist, most probably because I did not follow the wise counsel offered. Remaining errors are consequently my responsibility.

References

Bauer, L. and Nation, P. (1993) 'Word families'. *International Journal of Lexicography*, 64, 253–79.

Biber, D., Johansson, S., Leech, G., Conrad, S. and Finegan, E. (1999) *Longman Grammar of Spoken and Written English*. Harlow: Longman.

Biermann, A. (1997) *Great Ideas in Computer Science – A Gentle Introduction* (2nd edn). Cambridge, MA: MIT Press.

Brinton, D. M., Snow, M. A. and Wesche, M. H. (1989) *Content-Based Second Language Instruction*. New York: Harper and Row.

Byrnes, H., Weger-Guntharp, H. and Sprang, K. A. (eds) (2006) *Educating for Advanced Foreign Language Capacities: Constructs, Curriculum, Instruction, Assessment*. Washington, DC: Georgetown University Press.

Carroll, M. and M. Lambert (2006) 'Reorganizing principles of information structure in advanced L2s: French and German learners of English', in H. Byrnes, H. Weger-Guntharp and K. A. Sprang (eds), pp. 54–73.

Carter, R. and McCarthy, M. (eds) (1988) *Vocabulary and Language Teaching*. Harlow: Longman.

Coxhead, A. (2000) 'A new academic word list'. *TESOL Quarterly*, 34, 213–38.

Coxhead, A. and Nation, P. (2001) 'The specialised vocabulary of English for academic purposes', in J. Flowerdew and M. Peacock (eds), *Research Perspectives on English for Academic Purposes*. Cambridge: Cambridge University Press, pp. 252–67.

Dulay, H. and Burt, M. (1974) 'Natural sequences in child second language acquisition'. *Language Learning*, 24, 37–53.

Dunning, T. (1993) 'Accurate methods for the statistics of surprise and coincidence'. *Computational Linguistics*, 19, 61–74.

Ellis, N. and Sinclair, S. (1996) 'Working memory in the acquisition of vocabulary and syntax: Putting language in good order'. *The Quarterly Journal of Experimental Psychology*, 49A/1, 234–50.

Flowerdew, J. (1993) 'Concordancing as a tool in syllabus design'. *System*, 21, 231–44.

Ghadessy, M., Henry, A. and Roseberry, R. (eds) (2001) *Small Corpus Studies and ELT*. Amsterdam/Philadelphia: John Benjamins.

Halliday, M. A. K. (1956/1976) 'Grammatical categories in Modern Chinese'. *Transactions of the Philological Society* 1956, 180–202. Abridged and reprinted in G. Kress (ed.), *Halliday: System and Function*. Oxford: Oxford University Press, pp. 36–51.

Halliday, M. A. K. (1975) *Learning How to Mean: Explorations in the Development of Language*. London: Edward Arnold.

Halliday, M. A. K. (1978) *Language as Social Semiotic: The Social Interpretation of Language and Meaning*. London: Edward Arnold.

Halliday, M. A. K. (1989) *Spoken and Written Language*. Oxford: Oxford University Press.

Halliday, M. A. K. (1991a) 'Towards probabilistic interpretations', in E. Ventola (ed.), *Functional and Systemic Linguistics: Approaches and Uses*. Berlin: Mouton de Gruyter, pp. 39–61.

Halliday, M. A. K. (1991b) 'Corpus studies and probabilistic grammar', in K. Aijmer and B. Altenberg (eds), *English Corpus Linguistics*. Harlow: Longman, pp. 30–43.

Halliday, M. A. K. (1993) 'Towards a language-based theory of learning'. *Linguistics and Education*, 5, 93–116.

Halliday, M. A. K. and Hasan, R. (1985) *Language, Context and Text: Aspects of Language in a Social-semiotic Perspective*. Geelong, Australia: Deakin University Press.

Halliday, M. A. K. and James, Z. (1993) 'A quantitative study of polarity and primary tense in the English finite clause', in J. Sinclair, M. Hoey and B. Fox (eds), *Techniques of Description*. London: Routledge, pp. 32–66.

Halliday, M. A. K. and Martin, J. R. (1993) *Writing Science – Literacy and Discursive Power*. London: The Falmer Press.

Halliday, M. A. K. and Matthiessen, C. M. I. M. (2004) *An Introduction to Functional Grammar* (3rd edn). London: Arnold.

Hasan, R. (1979) 'On the notion of text', in J. Petöfi (ed.), *Text vs Sentence*. Hamburg: Helmut Burke Verlag, pp. 369–89.

Hirsh, D. and Nation, P. (1992) 'What vocabulary size is needed to read unsimplified texts for pleasure?' *Reading in a Foreign Language*, 8/2, 689–96.

Hoey, M. (2004) 'Lexical priming and the properties of text', in A. Partington, J. Morley and L. Haarman (eds), *Corpora and Discourse*. Bern: Peter Lang, pp. 385–412.

Hoey, M. (2005) *Lexical Priming*. London: Routledge.

Hyland, K. (1998) 'Boosting, hedging and the negotiation of academic knowledge'. *Text*, 18, 349–82.

Hyland, K. (1999) 'Disciplinary discourses: Writer stance in research articles', in C. Candlin and K. Hyland (eds), *Writing: Texts, Processes and Practices.* Harlow: Addison Wesley Longman, pp. 99–121.
Hyland, K. (2002a) 'What do they mean? Questions in academic writing'. *Text*, 22, 529–57.
Hyland, K. (2002b) 'Authority and invisibility: authorial identity in academic writing'. *Journal of Pragmatics*, 34, 1091–1112.
Krashen, S. (1985) *The Input Hypothesis.* London: Longman.
Larsen-Freeman, D. (1975) 'The acquisition of grammatical morphemes by adult ESL students'. *TESOL Quarterly*, 9, 409–19.
Leckie-Tarry, H. (1995) *Language and Context – A Functional Linguistic Theory of Register.* London: Pinter.
Lewis, M. (1993) *The Lexical Approach.* Hove: Language Teaching Publications.
Martin, J. R. (1984/2001) 'Language, register and genre', in F. Christie (ed.), *ECT 418 Children Writing: A Reader.* Geelong, Australia: Deakin University Press. Abridged and reprinted in A. Burns and C. Coffin (eds), *Analysing English in a Global Context.* London: Routledge, pp. 149–66.
Martin, J. R. (1989) *Factual Writing: Exploring and Challenging Social Reality.* Oxford: Oxford University Press.
Martin, J. R. (1992) *English Text: System and Structure.* Amsterdam/Philadelphia: John Benjamins.
Martin, J. R. (1997) 'Analysing genre: Functional parameters', in F. Christie and J. R. Martin (eds), *Genre and Institutions.* London: Cassell, pp. 3–39.
Martin, J. R. (1999) 'Modelling context', in M. Ghadessy (ed.), *Text and Context in Functional Linguistics.* Amsterdam/Philadelphia: John Benjamins, pp. 25–61.
Martin, J. R. and Veel, R. (eds) (1998) *Reading Science.* London: Routledge.
Matthiessen, C. M. I. M. (1993) 'Register in the round: diversity in a unified theory of register analysis', in M. Ghadessy (ed.), *Register Analysis: Theory and Practice.* London: Pinter, pp. 221–92.
Matthiessen, C. M. I. M. (forthcoming) 'Frequency profiles of some basic grammatical systems: an interim report', in G. Thompson and S. Hunston (eds), *System and Corpus – Exploring Connections.* London: Equinox.
McNamara, T. (2001) 'Language practice as social practice: challenges for research'. *Language Testing*, 18, 333–49.
Mohan, B. (1986) *Language and Content.* Reading, MA: Addison Wesley.
Moore, N. (2000) 'What is Computer English?', in S. Riley, S. Troudi and C. Coombe (eds), *Teaching, Learning and Technology.* Al Ain: TESOL Arabia, 29–41.
Nation, I. S. P. (1999) *Learning Vocabulary in Another Language.* English Language Institute Occasional Publication no. 19. Wellington, New Zealand: Victoria University of Wellington.
Nation, I. S. P. (2002) *VocabProfile.* Software available online at http://www.vuw.ac.nz/lals/staff/Paul_Nation; accessed on 20 April 2004.
Nation, P. and Waring, R. (1997) 'Vocabulary size, text coverage and word lists', in N. Schmitt and M. McCarthy (eds), *Vocabulary Description, Acquisition and Pedagogy.* Cambridge: Cambridge University Press, pp. 6–19.
Nattinger, J. and DeCarrico, J. (1992) *Lexical Phrases and Language Teaching.* Oxford: Oxford University Press.
Nesbitt, C. and Plum, G. (1988) 'Probabilities in a systemic-functional grammar: the clause complex in English', in R. Fawcett and D. Young (eds), *New Developments in Systemic Linguistics, Vol. 2: Theory and Application.* London: Pinter, pp. 6–38.

Painter, C. (1989) *Learning the Mother Tongue.* Oxford: Oxford University Press.

Paribakht, T. S. and Wesche, M. (2006) 'Lexical inferencing in L1 and L2: implications for vocabulary instruction and learning at advanced levels', in H. Byrnes, H. Weger-Guntharp and K. A. Sprang (eds), pp. 118–35.

Pawley, A. and F. Syder, F. (1983) 'Two puzzles for linguistic theory: nativelike selection and nativelike fluency', in J. Richards and R. Schmidt (eds), *Language and Communication.* Harlow: Longman, pp. 191–226.

Peters, A. M. (1983) *The Units of Language Acquisition.* Cambridge: Cambridge University Press.

Roe, P. J. (1995) *Aston Text Analyser.* Software available from Language Studies Unit, Aston University.

Schleppegrell, M. J. (2001) 'Linguistic features of the language of schooling'. *Linguistics and Education,* 14, 431–59.

Schleppegrell, M. J., Achugar, M. and Oteiza, T. (2004) 'The grammar of history: Enhancing content-based instruction through a functional focus on language'. *TESOL Quarterly,* 38, 67–93.

Schmidt, R. (1990) 'The role of consciousness in second language learning'. *Applied Linguistics,* 11, 129–58.

Schmidt, R. (1993) 'Awareness and second language acquisition'. *Annual Review of Applied Linguistics,* 13, 206–26.

Scott, M. (2005) *Wordsmith Tools v4.1.* Software licensed by Oxford University Press.

Sinclair, J. (1991) *Corpus, Concordance, Collocation.* Oxford: Oxford University Press.

Sinclair, J. (ed.) (1987) *Looking Up.* London: Collins.

Sinclair, J. and Renouf, A. (1988) A lexical syllabus for language learning', in R. Carter and M. McCarthy (eds), *Vocabulary and Language Teaching.* Harlow: Longman, pp. 140–60.

Stutterheim, von C. and Carroll, M. (2006) 'The impact of grammatical temporal categories on ultimate attainment in L2 learning', in H. Byrnes, H. Weger-Guntharp and K. A. Sprang (eds), pp. 40–53.

Tadros, A. (1989) 'Predictive categories in university textbooks'. *English for Specific Purposes,* 8, 17–31.

Thompson, G. (2001) 'Interaction in academic writing: learning to argue with the reader'. *Applied Linguistics,* 22, 58–78.

Ventola, E. (1987) *The Structure of Social Interaction: A Systemic Approach to the Semiotics of Service Encounters.* London: Pinter.

Ventola, E. (1996) 'Packing and unpacking of information in academic texts', in E. Ventola and A. Mauranen (eds), *Academic Writing: Intercultural and Textual Issues.* Amsterdam/Philadelphia: John Benjamins, pp. 153–94.

Vygotsky, L. (1966/1991) 'Genesis of the higher mental functions', in A. Leontyev, A. Luria and A. Smirnoff (eds), *Psychological Research in the USSR* vol. 1. Moscow: Progress Publishers. Reprinted in P. Light, S. Sheldon and M. Woodhead (eds), *Child Development in Social Context.* London: Routledge, pp. 32–41.

Ward, J. (1999) 'How large a vocabulary do EAP engineering students need?', *Reading in a Foreign Language,* 12, 309–23.

Willis, D. (1990) *The Lexical Syllabus.* London: Collins Cobuild.

Willis, D. (2003) *Rules, Patterns and Words.* Cambridge: Cambridge University Press.

Wray, A. (2002) *Formulaic Language and the Lexicon.* Cambridge: Cambridge University Press.

Index